SpringerBriefs in Statistics

For further volumes:
http://www.springer.com/series/8921

Bronius Grigelionis

Student's *t*-Distribution and Related Stochastic Processes

Springer

Bronius Grigelionis
University of Vilnius
Vilnius
Lithuania

ISSN 2191-544X ISSN 2191-5458 (electronic)
ISBN 978-3-642-31145-1 ISBN 978-3-642-31146-8 (eBook)
DOI 10.1007/978-3-642-31146-8
Springer Heidelberg New York Dordrecht London

Library of Congress Control Number: 2012945743

Printed on acid-free paper

Springer is part of Springer Science+Business Media (www.springer.com)

To my family

Preface

Stochastic processes with heavy-tailed marginal distributions, including Student's t-distribution, are used commonly for modelling in communication networks, econometrics, insurance, logarithmic stock returns and stochastic volatility in finance, electric activity of neurons, turbulence, etc.

The aim of this short book is the survey of recent result on the Student–Lévy processes as a subclass of Thorin subordinated Gaussian–Lévy processes. Criteria of self-decomposability of such processes are discussed in detail and related Ornstein–Uhlenbeck-type processes are constructed.

The univariate Student diffusion processes are considered in the framework of the H-diffusions, i.e., stationary ergodic diffusions with the predetermined marginal distribution H. Asymptotic distributions of the normalised extreme values of these diffusions are given. Special attention is paid to the statistically tractable case of the Kolmogorov–Pearson diffusions.

Using the independently scattered random measures, defined by means of the bivariate Student–Lévy processes, strictly stationary Student processes with the arbitrary correlation function are defined. Further, via the Lamperti's transform, the self-similar Student–Lamperti processes are introduced.

As a promising direction for future work in constructing and investigating of new multivariate Student–Lévy-type processes, the notion of Lévy copulas and the related analogue of Sklar's theorem is briefly explained.

Statistical inference problems as well as general studentised statistics and self-normalised processes are not considered at all. List of references is far from to be complete.

The author is grateful to the colleagues Algimantas Bikelis, Kęstutis Kubilius, Kazimieras Padvelskis and Pranas Vaitkus for friendly support and Rimantė Baltutytė for excellent typing.

Vilnius, September 2011 Bronius Grigelionis

Contents

Abstract and Keywords

Abstract This brief monograph contains a deep study of infinite divisibility and self-decomposability properties of central and non-central Student's distributions, represented as variance and mean-variance mixtures of multivariate Gaussian distributions with the reciprocal gamma mixing distribution, respectively. These results permit to define and analyse Student–Lévy processes as Thorin subordinated Gaussian–Lévy processes. Analogously, Student–Ornstein–Uhlenbeck-type processes are described. A wide class of one-dimensional strictly stationary diffusions with the Student's t marginal distribution is defined as the unique weak solution for the stochastic differential equation. Extreme value theory for such diffusions is developed. A flexible and statistically tractable Kolmogorov–Pearson diffusions are also described. Using the independently scattered random measures, generated by the bivariate centered Student–Lévy process, and stochastic integration theory with respect to them, it is defined as an univariate strictly stationary process with the centered Student's t marginals and the arbitrary correlation structure. As a promising direction for future work in constructing and analysing of new multivariate Student–Lévy-type processes, the notion of Lévy copulas and the related analogue of Sklar's theorem is explained.

Keywords Bessel function · Gaussian Lévy process · H-diffusion · Self-decomposability · Stationary Student process · Student–Lévy process · Student's t-distribution · Thorin subordinator · Tweedie class

Chapter 1
Introduction

Considering a sample of independent observations $X_1, \dots, X_n$ from the normal population with mean α and variance σ^2 for testing the null hypothesis $H_0 : \alpha = \alpha_0$ against the alternative $H_1 : \alpha = \alpha_1$, Gosset ("Student") in 1908 [1] suggested the test statistic

$$t_n = \frac{\sqrt{n}(\bar{X}_n - \alpha_0)}{s_n}, \quad n \geq 2,$$

where $\bar{X}_n = \frac{1}{n}\sum_{j=1}^{n} X_j$, $s_n^2 = \frac{1}{n-1}\sum_{j=1}^{n}(X_j - \bar{X}_n)^2$. He derived that the distribution law

$$\mathscr{L}(t_n) = T_1(n-1, 1, 0),$$

where $T_1(\nu, \sigma^2, \alpha)$ denotes the univariate Student's t-distribution with $\nu > 0$ degrees of freedom, a scaling parameter $\sigma^2 > 0$ and a location parameter $\alpha \in R^1$, defined by its probability density function (pdf for short) $f_{\nu,\sigma^2}(x - \alpha)$, where

$$f_{\nu,\sigma^2}(x) = \frac{\Gamma(\frac{\nu+1}{2})}{\sqrt{\pi\nu}\sigma\Gamma(\frac{\nu}{2})}\left[1 + \frac{1}{\nu}\left(\frac{x}{\sigma}\right)^2\right]^{-\frac{\nu+1}{2}}, \quad x \in R^1,$$

and $\Gamma(z)$ is the Euler's gamma function (see [2]).

Having in mind that the statistics $\bar{X}_n$ and s_n^2 are independent, $\mathscr{L}(\bar{X}_n) = N(\alpha_0, \frac{\sigma^2}{n})$ and $\mathscr{L}(\sigma^{-2}s_n^2) = \Gamma_{\frac{n-1}{2},\frac{n-1}{2}}$, we easily find that

$$f_{\nu,\sigma^2}(x) = \int_0^\infty \frac{1}{\sqrt{2\pi y\sigma^2}} e^{-\frac{1}{2y}\left(\frac{x}{\sigma}\right)^2} h_\nu(y)\mathrm{d}y,$$

where $\Gamma_{\beta,\gamma}$ is the gamma distribution with the pdf

B. Grigelionis, *Student's* t-*Distribution and Related Stochastic Processes*, SpringerBriefs in Statistics, DOI: 10.1007/978-3-642-31146-8_1,

$$p_{\beta,\gamma}(x) = \begin{cases} \frac{\beta^\gamma}{\Gamma(\gamma)} x^{\gamma-1} e^{-\beta x}, & \text{if } x > 0, \\ 0, \text{x} \leq 0, \end{cases}$$

and

$$h_\nu(y) = \frac{(\frac{\nu}{2})^{\frac{\nu}{2}}}{\Gamma(\frac{\nu}{2})} y^{-\frac{\nu}{2}-1} e^{-\frac{\nu}{2y}}, \quad y > 0,$$

is the pdf of the inverse (reciprocal) gamma distribution $I\Gamma_{\frac{\nu}{2},\frac{\nu}{2}}$.

In 1931 [3] Fisher introduced the univariate noncentral t-distribution with the pdf $f_{\nu,\sigma^2,a}(x-\alpha)$ as a mean–variance mixture of normal distributions with the inverse gamma mixing distribution, i.e.

$$f_{\nu,\sigma^2,a}(x) = \int_0^\infty \frac{1}{\sqrt{2\pi y \sigma^2}} e^{-\frac{1}{2y}\left(\frac{x-ay}{\sigma}\right)^2} h_\nu(y)\mathrm{d}y$$

$$= \frac{(\frac{\nu}{2})^{\frac{\nu}{2}}}{\Gamma(\frac{\nu}{2})} \frac{2\exp\{\frac{xa}{\sigma^2}\}}{\sqrt{2\pi\sigma^2}} \left(\frac{a^2}{\nu\sigma^2 + x^2}\right)^{\frac{\nu+d}{4}} K_{\frac{\nu+1}{2}}\left(\sigma^{-2}[a^2(\nu\sigma^2 + x^2)]^{\frac{1}{2}}\right), \quad x \in R^1,$$

where $K_\nu(z)$ is the modified Bessel function of the third kind (see Appendix).

There are unlimited possibilities to introduce classes of multivariate extensions of Student's t-distributions with the univariate Student's marginals. An excellent survey of such useful generalizations are given by Kotz and Nadarajah in [4] (see also [5]). Further we shall mainly restrict ourselves to the cases of variance mixtures and mean–variance mixtures of multivariate Gaussian distributions with the inverse gamma mixing distribution h_ν.

Let

$$g_{a,\Sigma}(x) = \frac{1}{\sqrt{|\Sigma|}(2\pi)^{\frac{d}{2}}} \exp\left\{-\frac{1}{2}\langle (x-a)\Sigma^{-1}, x-a\rangle\right\}, \quad x \in R^d,$$

be a Gaussian pdf, where $a \in R^d$, Σ is a symmetric positive definite $d \times d$ matrix, $|\Sigma| := \det \Sigma$, $\langle \cdot, \cdot \rangle$ signs the scalar product in R^d.

Definition 1.1 We say that $T_d(\nu, \Sigma, \alpha)$ is a multivariate Student's t-distribution with $\nu > 0$ degrees of freedom, a scaling matrix Σ and a location vector $\alpha \in R^d$, if its pdf is $f_{\nu,\Sigma}(x-\alpha)$, $x \in R^d$, where

$$f_{\nu,\Sigma}(x) = \int_0^\infty g_{0,u\Sigma}(x) h_\nu(u)\mathrm{d}u$$

$$= \frac{(\frac{\nu}{2})^{\frac{\nu}{2}}}{\Gamma(\frac{\nu}{2})\sqrt{|\Sigma|}(2\pi)^{\frac{d}{2}}} \int_0^\infty u^{-\frac{d}{2}} e^{-\frac{1}{2u}\langle x\Sigma^{-1},x\rangle} u^{-\frac{\nu}{2}-1} e^{-\frac{\nu}{2u}} \mathrm{d}u$$

$$= \frac{\Gamma(\frac{\nu+d}{2})}{(\nu\pi)^{\frac{d}{2}}\Gamma(\frac{\nu}{2})\sqrt{|\Sigma|}} \left(1 + \frac{\langle x\Sigma^{-1}, x\rangle}{\nu}\right)^{-\frac{\nu+d}{2}}, \quad x \in R^d. \tag{1.1}$$

Definition 1.2 We say that $T_d(\nu, \Sigma, \alpha, a)$ is a noncentral multivariate Student's t-distribution with $\nu > 0$ degrees of freedom, a scaling matrix Σ, a location vector $\alpha \in R^d$, and a noncentrality vector $a \in R^d \backslash \{0\}$, if its pdf is $f_{\nu,\Sigma,a}(x - \alpha)$, $x \in R^d$, where

$$f_{\nu,\Sigma,a}(x) = \int_0^\infty g_{ua,u\Sigma}(x) h_\nu(u) \mathrm{d}u$$

$$= \frac{(\frac{\nu}{2})^{\frac{\nu}{2}}}{\Gamma(\frac{\nu}{2})\sqrt{|\Sigma|}(2\pi)^{\frac{d}{2}}} \int_0^\infty u^{-\frac{d}{2}} e^{-\frac{1}{2u}\langle (x-ua)\Sigma^{-1},x-ua\rangle} u^{-\frac{\nu}{2}-1} e^{-\frac{\nu}{2u}} \mathrm{d}u$$

$$= \frac{2(\frac{\nu}{2})^{\frac{\nu}{2}} \exp\{\langle x\Sigma^{-1}, a\rangle\}}{\Gamma(\frac{\nu}{2})\sqrt{|\Sigma|}(2\pi)^{\frac{d}{2}}} \left(\frac{\langle a\Sigma^{-1}, a\rangle}{\nu + \langle x\Sigma^{-1}, x\rangle}\right)^{\frac{\nu+d}{4}}$$

$$\times K_{\frac{\nu+d}{2}}\left(\left[\langle a\Sigma^{-1}, a\rangle(\nu + \langle x\Sigma^{-1}, x\rangle)\right]^{\frac{1}{2}}\right), \quad x \in R^d. \tag{1.2}$$

Relating Student's t-distributions to the Lévy processes or to the Ornstein–Uhlenbeck type processes the crucial role are paid the properties of infinite divisibility or self-decomposability (for used terminology see [6] or Chap. 3 below). Intensive studies of new criteria for such properties began in 1970s of last century (see, e.g., [7–11]). In this sense two results are of the key importance.

From the one hand, in 1976 [7] Grosswald proved that the univariate Student's t-distribution of any degree of freedom is infinitely divisible, deriving the following formula:

$$K_{\nu-1}(x) = xK_\nu(x) \int_0^\infty \frac{g_\nu(u)}{x^2 + u} du, \quad \nu \geq -1, x > 0, \tag{1.3}$$

where

$$g_\nu(x) = 2\left\{\pi^2 x(J_\nu^2(\sqrt{x}) + Y_\nu^2(\sqrt{x}))\right\}^{-1}, \quad x > 0,$$

$J_\nu(z)$ and $Y_\nu(z)$ are the Bessel functions of the first kind and the second kind, respectively (see Appendix).

From the second hand, in 1977 [11] Thorin defined the class of generalized gamma convolutions (GGC or $T_1(R_+)$ for short) as the minimal class of probability distributions on $R_+ := [0, \infty)$ containing all gamma distributions, closed under convolutions and weak limits. He proved that $\tau \in T_1(R_+)$ if and only if the Laplace transform of τ has the form:

$$\int_0^\infty e^{-\theta u}\tau(du) = \exp\left\{-\beta_0\theta + \int_0^\infty (e^{-\theta u} - 1)\frac{1}{u}\int_0^\infty e^{-vu}Q_1(dv)du\right\}$$
$$= \exp\left\{-\beta_0\theta + \int_0^\infty \log\left(\frac{v}{\theta + v}\right)Q_1(dv)\right\}, \quad \theta > 0, \tag{1.4}$$

where $\beta_0 \geq 0$ and Q_1 is a Radon measure on R_+ such that $Q_1(\{0\}) = 0$,

$$\int_0^1 \log\left(\frac{1}{u}\right)Q_1(du) < \infty \quad \text{and} \quad \int_1^\infty u^{-1}Q_1(du) < \infty.$$

(see [11–14]).

All distributions in $T_1(R_+)$ are self-decomposable.

For example, generalized inverse (GIG for short) Gaussian distributions, defined by the pdf

$$gig(x; \lambda, \chi, \psi) = \frac{(\frac{\psi}{\chi})^{\frac{\lambda}{2}}}{2K_\lambda(\sqrt{\chi\psi})}x^{\lambda-1}\exp\left\{-\frac{1}{2}(\chi x^{-1} + \psi x)\right\}, \quad x > 0,$$

where $\lambda \in R^1$, $(\chi, \psi) \in \Theta_\lambda$,

$$\Theta_\lambda = \begin{cases} \{(\chi, \psi) : \chi \geq 0, \psi > 0\}, & \text{if } \lambda > 0, \\ \{(\chi, \psi) : \chi > 0, \psi > 0\}, & \text{if } \lambda = 0, \\ \{(\chi, \psi) : \chi > 0, \psi \geq 0\}, & \text{if } \lambda < 0, \end{cases}$$

are in class $T_1(R_+)$, because it is easy to check that

$$\int_0^\infty e^{-\theta u}gig(u; \lambda, \chi, \psi)du = \left(\frac{\psi}{\psi + 2\theta}\right)^{\frac{\lambda}{2}}\frac{K_\lambda(\sqrt{\chi(\psi + 2\theta)})}{K_\lambda(\sqrt{\chi\psi})}$$

and, using Grosswald's formula (3), to derive that

$$\int_0^\infty e^{-\theta u}gig(u; \lambda, \chi, \psi)du = \exp\left\{\int_0^\infty (e^{-\theta u} - 1)\frac{1}{u}\int_0^\infty e^{-vu}Q_1(dv)du\right\}, \quad \theta > 0,$$

with $Q_1(\mathrm{d}v) = \lambda\varepsilon_{\frac{\psi}{2}}(\mathrm{d}v)$ for $\chi = 0$, $\lambda > 0$, and

$$Q_1(\mathrm{d}v) = \max(0,\lambda)\varepsilon_{\frac{\psi}{2}}(\mathrm{d}v) + 1_{(\frac{\psi}{2},\infty)}\chi g_{|\lambda|}(\chi(2t-\psi))\mathrm{d}t \tag{1.5}$$

for $\chi > 0$ (see, e.g., [15]). Here ε_Δ is the Dirac measure and 1_B is the indicator function .

Observe that

$$gig(x;\lambda,0,\psi) = \frac{(\frac{\psi}{2})^\lambda}{\Gamma(\lambda)}x^{\lambda-1}e^{-\frac{\psi}{2}x}, \quad \lambda > 0, \quad \psi > 0,$$

and

$$gig(x;\lambda,\chi,0) = \frac{1}{\Gamma(-\lambda)}\left(\frac{2}{\chi}\right)^\lambda x^{\lambda-1}e^{-\frac{\chi}{2}x^{-1}}, \quad \lambda < 0, \quad \chi > 0.$$

Thus, the mixing distribution h_ν in (1.1) and (1.2) is from the Thorin class.

Infinite divisible distributions on R_+ correspond one-to-one with the Lévy processes, starting at zero and having nondecreasing trajectories called the subordinators. In particular, Thorin distributions define the class of Thorin subordinators.

At last, exploiting the Bochner's idea of subordination and using the multivariate Gaussian Lévy processes as subordinands and GIG subordinators, we shall obtain the important class of generalized hyperbolic processes, introduced by Barndorff-Nielsen (see, e.g., [15]), which contain the Student-Lévy processes, generated by the mixtures (1.1) and (1.2).

These types of stochastic processes as well as stochastic processes with heavy-tailed marginals like Student's t-distributions are commonly used for modeling in various fields of applications (see, e.g., [16–22] and references therein).

In Chap. 2 of this brief monograph there are presented asymptotics of Student's t pdf as a degree of freedom ν or arguments $|x|$ tend to infinity. In the one-dimensional case asymptotic distributions for extremal and record values in i.i.d. sequences of random variables with common Student's t-distribution are described.

In Chap. 3 via Lévy-Itô decomposition the structure of d-dimensional Lévy processes is explained including the celebrated Lévy-Khinchine formula for characteristic functions of infinitely divisible laws in R^d. Criteria of their self-decomposability are derived. Extending the Thorin class $T_1(R_+)$ and related Lévy subordinators the scale of Thorin classes $T_\varkappa(R_+)$, $0 < \varkappa \leq \infty$, is defined and characterized as generalized convolutions of the famous Tweedie distributions.

Subordination of Lévy processes as a tool for construction and investigation of new Lévy processes with the desirable distributional properties is also discussed.

In Chap. 4 there are characterized the Thorin subordinated Gaussian-Lévy processes, including Student-Lévy processes. Criteria of their self-decomposability are derived.

In Chap. 5 the Student Ornstein–Uhlenbeck type processes are studied.

In Chap. 6 the strictly stationary regular positive recurrent diffusion processes on an open interval $(l, r) \subseteq R^1$ with inaccessible end points and predetermined 1D distributions H are considered and named H-diffusions. The class of Student diffusions as H-diffusions on R^1 with H equal to the univariate Student distribution is investigated in detail. Asymptotic distributions of extreme values of H-diffusions, including the Student ones, are derived. Conditions of vague convergence of time normalized point measures of ε-upcrossings of such diffusions to the Poisson point measures are discussed.

As the flexible and statistically tractable stochastic processes, the Kolmogorov-Pearson diffusions are described.

In the final Chap. 7 it is presented extended Isserlis theorem, giving formulas for mixed moments of mixtures of Gaussian distributions and, as a special case, for Student's distributions.

Using the independently scattered random measures there are constructed strictly stationary real stochastic processes $X = \{X_t, t \in R^1\}$ such that

$$\mathscr{L}(X_t) \equiv T_1(\nu, \sigma^2, \alpha), \quad \nu > 2$$

and related self-similar processes, obtained by means of the Lamperti's transform.

Following [23], as a promising direction for the future work a notion of Lévy copulas and analog of well-known Sklar's theorem are explained.

Appendix contains used notions and formulas from the theory of Bessel functions (see, e.g., [24, 25]).

References

1. Student: On the probable error of mean. Biometrika. **6**, 1–25 (1908)
2. Fisher, R.A.: Applications of Student's distribution. Metron **5**, 109–112 (1925)
3. Fisher, R.A.: Introduction to Table of Hh functions. In: British Association Mathematical Tables, vol. 1, pp. 26–35. British Association, London, (1931)
4. Kotz, S., Nadarajah, S.: Multivariate *t* Distributions and Their Applications. Cambridge University Press, Cambridge (2004)
5. Johnson, N.L., Kotz, S.: Distributions in Statistics: Continuous Univariate Distributions. Wiley, New York (1972)
6. Sato, K.: Lévy Processes and Infinitely Divisible Distributions. Cambridge University Press, Cambridge (1999)
7. Grosswald, E.: The Student *t*-distribution of any degree of freedom is infinitely divisible. Z. Wahrscheinlichkeitstheor. verw. Geb. **36**, 103–109 (1976)
8. Halgreen, C.: Self-decomposability of the generalized inverse Gaussian and hyperbolic distributions. Z. Wahrscheinlichkeitstheor. verw. Geb. **47**, 13–17 (1979)
9. Ismail, M.E.H., Kelker, D.H.: The Bessesl polynomials and the Student *t*-distribution. SIAM J. Math. Anal. **7**, 82–91 (1976)
10. Kelker, D.: Infinite divisibility and variance mixtures of the normal distribution. Ann. Math. Statist. **42**, 802–808 (1971)
11. Thorin, O.: On the infinite divisibility of lognormal distribution. Scand. Actuar. J. **3**, 121–148 (1977)

12. Bondesson, L.: Generalized gamma convolutions and related classes of distributions and densities. Lecture Notes in Statistics, vol. 76. Springer-Verlag, Berlin (1992)
13. Grigelionis, B.: Extending the Thorin class. Lith. Math. J. **51**(2), 194–206 (2011)
14. James, L.F., Roynette, B., Yor, M.: Generalized gamma convolutions, Dirichlet means, Thorin measures, with explicit examples. Probab. Surv. **5**, 346–415 (2008)
15. Barndorff-Nielsen, O.E.: Exponentially decreasing distributions for the logarithm of particle size. Proc. Royal Soc. Lond. A **353**, 401–419 (1977)
16. Barndorff-Niesen, O.E., Shephard, N.: Non-Gaussian Ornstein Uhlenbeck-based models and some of their use in financial econometrics. J. R. Statist. Soc. B. **63**, 167–241 (2001)
17. Cont, R., Tankov, P.: Financial Modelling with Jump Processes. Chapman & Hall/CRC, Boca Raton (2004)
18. Heyde, C.C., Leonenko, N.N: Student processes, Adv. Appl. Prob. **37**, 342–365 (2005)
19. Hurst, S.R., Platen, E., Rachev, S.R.: Subordinated Markov models: a comparison. Finan. Eng. Jpn. Markets **4**, 97–124 (1997)
20. McNeil, A.J., Frey, R., Embrechts, P.: Quantitive Risk Management: Concepts, Techniques, and Tools. Princeton series in, finance (2005)
21. Schoutens, W.: Lévy Processes in Finance: Pricing Financial Derivatives. Wiley, New York (2003)
22. Woyczynski, W.A: Burgers-KPZ turbulence. Lecture Notes Mathematics, vol. 1700, Springer, Berlin (1998)
23. Kallsen, J., Tankov, P.: Characterization of dependence of multidimensional Lévy processes using Lévy copulas. J. Multivar. Anal. **97**, 1551–1572 (2006)
24. Abramowitz, M., Stegun, I. (eds.): Handbook of Mathematical Functions. Dover, New York (1968)
25. Watson, G.N.: Treatise on the Theory of Bessel Functions. Cambridge University Press, Cambridge (1958)

Chapter 2
Asymptotics

2.1 Asymptotic Behavior of Student's Pdf

Proposition 2.1 *For each* $x \in R^d$, *as* $\nu \to \infty$,

$$f_{\nu,\Sigma,a}(x) \to g_{a,\Sigma}(x). \tag{2.1}$$

Proof Let $a = 0$. Using the well-known formula that

$$\Gamma(z) = \sqrt{\frac{2\pi}{z}} e^{-z} z^z \left(1 + O\left(\frac{1}{z}\right)\right), \quad \text{as} \quad z \to \infty, \tag{2.2}$$

we find that, as $\nu \to \infty$,

$$\frac{\Gamma(\frac{\nu+d}{2})}{(\nu\pi)^{\frac{d}{2}}\Gamma(\frac{\nu}{2})} \sim \frac{\sqrt{\frac{4\pi}{\nu+d}} e^{-\frac{\nu+d}{2}} (\frac{\nu+d}{2})^{\frac{\nu+d}{2}}}{(\nu\pi)^{\frac{d}{2}} \sqrt{\frac{4\pi}{\nu}} e^{-\frac{\nu}{2}} (\frac{\nu}{2})^{\frac{\nu}{2}}} \to \frac{1}{(2\pi)^{\frac{d}{2}}} \tag{2.3}$$

and, obviously,

$$\left(1 + \frac{\langle x\Sigma^{-1}, x\rangle}{\nu}\right)^{-\frac{\nu+d}{2}} \to e^{-\frac{1}{2}\langle x\Sigma^{-1}, x\rangle}. \tag{2.4}$$

Here and below "$\sim$" is the equivalence sign.

The statement (2.1) with $a = 0$ follows from (1.1), (2.2), (2.3) and (2.4).

Let now $a \neq 0$ and

$$y_\nu = \frac{2}{\nu + d}\left[\langle a\Sigma^{-1}, a\rangle(\nu + \langle x\Sigma^{-1}, x\rangle)\right]^{\frac{1}{2}}.$$

B. Grigelionis, *Student's t-Distribution and Related Stochastic Processes*,
SpringerBriefs in Statistics, DOI: 10.1007/978-3-642-31146-8_2,

Because, as $\nu \to \infty$, uniformly in y (see Appendix)

$$K_\nu(\nu y) \sim \sqrt{\frac{\pi}{2\nu}} \frac{\exp\{-\nu\sqrt{1+y^2}\}}{(1+y^2)^{\frac{1}{4}}} \left(\frac{y}{1+\sqrt{1+y^2}}\right)^{-\nu}$$

and

$$\sqrt{1+y_\nu^2} \sim 1 + \frac{1}{2} y_\nu^2,$$

we shall have that

$$\begin{aligned}
&K_{\frac{\nu+d}{2}}\left(\left[\langle a\Sigma^{-1}, a\rangle(\nu + \langle x\Sigma^{-1}, x\rangle)\right]^{\frac{1}{2}}\right) = K_{\frac{\nu+d}{2}}\left(\frac{\nu+d}{2} y_\nu\right) \\
&\sim \sqrt{\frac{\pi}{\nu+d}} \exp\left\{-\frac{\nu+d}{2}\left(1+\frac{1}{2}y_\nu^2\right)\right\}\left(\frac{y_\nu}{2+\frac{1}{2}y_\nu^2}\right)^{-\frac{\nu+d}{2}} \\
&\sim \sqrt{\frac{\pi}{\nu+d}} e^{-\frac{\nu+d}{2}} \exp\left\{-\frac{1}{\nu+d}\langle a\Sigma^{-1}, a\rangle\left(\nu + \langle x\Sigma^{-1}, x\rangle\right)\right\}\left(\frac{y_\nu}{2+\frac{1}{2}y_\nu^2}\right)^{-\frac{\nu+d}{2}}.
\end{aligned} \tag{2.5}$$

From (1.2) and (2.5) we elementarily find that

$$\begin{aligned}
f_{\nu,\Sigma,a}(x) &\sim \frac{(\frac{\nu}{2})^{\frac{\nu}{2}}}{\Gamma(\frac{\nu}{2})} \frac{2\exp\{\langle x\Sigma^{-1}, a\rangle\}}{(2\pi)^{\frac{d}{2}}\sqrt{|\Sigma|}} \left(\frac{\langle a\Sigma^{-1}, a\rangle}{\nu_+ \langle x\Sigma^{-1}, x}\right)^{\frac{\nu+d}{4}} \sqrt{\frac{\pi}{\nu+d}} e^{-\frac{\nu+d}{2}} \\
&\quad \times \exp\left\{-\frac{1}{\nu+d}\langle a\Sigma^{-1}, a\rangle\left(\nu + \langle x\Sigma^{-1}, x\rangle\right)\right\}\left(\frac{y_\nu}{2+\frac{1}{2}y_\nu^2}\right)^{-\frac{\nu+d}{2}} \\
&\sim \frac{\exp\{\langle x\Sigma^{-1}, a\rangle\}}{(2\pi)^{\frac{d}{2}}\sqrt{|\Sigma|}} e^{-\langle a\Sigma^{-1}, a\rangle} e^{-\frac{d}{2}} \left(\frac{\nu + \langle x\Sigma^{-1}, x\rangle}{2+\frac{1}{2}y_\nu^2}\right)^{-\frac{\nu+d}{2}} \\
&\sim \frac{\exp\{\langle x\Sigma^{-1}, a\rangle\}}{(2\pi)^{\frac{d}{2}}\sqrt{|\Sigma|}} e^{-\langle a\Sigma^{-1}, a\rangle} e^{-\frac{d}{2}} \\
&\quad \times \exp\left\{-\frac{1}{2}(\langle x\Sigma^{-1}, x\rangle - d)\right\}\left(1+\frac{1}{4}y_\nu^2\right)^{\frac{\nu+d}{2}}.
\end{aligned} \tag{2.6}$$

But

$$\left(1+\frac{1}{4}y_\nu^2\right)^{\frac{\nu+d}{2}} = \left(1+\frac{1}{(\nu+d)^2}\left[\langle a\Sigma^{-1}, a\rangle\left(\nu + \langle x\Sigma^{-1}, x\rangle\right)\right]\right)^{\frac{\nu+d}{2}}$$

$$\to \exp\left\{\frac{1}{2}\langle a\Sigma^{-1}, a\rangle\right\}. \tag{2.7}$$

Thus, (2.6) and (2.7) imply that, for each $x \in R^d$, as $\nu \to \infty$,

$$f_{\nu,\Sigma,a}(x) \to \frac{\exp\left\{\langle x\Sigma^{-1}, a\rangle\right\}}{(2\pi)^{\frac{d}{2}}\sqrt{|\Sigma|}} \exp\left\{-\frac{1}{2}\left(\langle a\Sigma^{-1}, a\rangle + \langle x\Sigma^{-1}, x\rangle\right)\right\} = g_{a,\Sigma}(x). \quad \square$$

Proposition 2.2 *For each fixed $x \in R^d$ and $\nu > 0$, as $|a| \to 0$,*

$$f_{\nu,\Sigma,a}(x) \to f_{\nu,\Sigma}(x).$$

Proof Indeed, as $|a| \to 0$,

$$K_{\frac{\nu+d}{2}}\left(\left[\langle a\Sigma^{-1}, a\rangle(\nu + \langle x\Sigma^{-1}, x\rangle)\right]^{\frac{1}{2}}\right)$$
$$\sim \Gamma\left(\frac{\nu+d}{2}\right) 2^{\frac{\nu+d}{2}-1}\left[\langle a\Sigma^{-1}, a\rangle(\nu + \langle x\Sigma^{-1}, x\rangle)\right]^{-\frac{\nu+d}{4}}$$

(see Appendix) and, having in mind formulas (1.1), (1.2),

$$f_{\nu,\Sigma,a}(x) \to \frac{(\frac{\nu}{2})^{\frac{\nu}{2}}}{\Gamma(\frac{\nu}{2})} \frac{2^{\frac{\nu+d}{2}}\Gamma(\frac{\nu+d}{2})}{(2\pi)^{\frac{d}{2}}\sqrt{|\Sigma|}} \left(\nu + \langle x\Sigma^{-1}, x\rangle\right)^{-\frac{\nu+d}{2}} = f_{\nu,\Sigma}(x). \qquad \square$$

Proposition 2.3 *(i) As $|x| \to \infty$,*

$$f_{\nu,\Sigma}(x) \sim c_{\nu,\Sigma}\left(\langle x\Sigma^{-1}, x\rangle\right)^{-\frac{\nu+d}{2}},$$

where

$$c_{\nu,\Sigma} = \frac{\Gamma\left(\frac{d+\nu}{2}\right)}{\pi^{\frac{d}{2}}\Gamma(\frac{\nu}{2})\sqrt{|\Sigma|}}.$$

(ii) As $|x| \to \infty$, $a \neq 0$,

$$f_{\nu,\Sigma,a}(x) \sim c_{\nu,\Sigma,a}\left(\langle x\Sigma^{-1}, x\rangle\right)^{-\frac{\nu+d+1}{4}}$$

$$\times \exp\left\{-\left[\langle a\Sigma^{-1}, a\rangle\langle x\Sigma^{-1}, x\rangle\right]^{\frac{1}{2}} + \langle x\Sigma^{-1}, a\rangle\right\},$$

where

$$c_{\nu,\Sigma,a} = \frac{(\frac{\nu}{2})^{\frac{\nu}{2}}\left(\langle a\Sigma^{-1}, a\rangle\right)^{\frac{\nu+d+1}{4}}}{\Gamma(\frac{\nu}{2})(2\pi)^{\frac{d-1}{2}}\sqrt{|\Sigma|}}.$$

Proof (i) Obviously follows from (1.1).
(ii) Because, as $|x| \to \infty$,

$$K_{\frac{\nu+d}{2}}\left(\left[\langle a\Sigma^{-1}, a\rangle\left(\nu + \langle x\Sigma^{-1}, x\rangle\right)\right]^{\frac{1}{2}}\right)$$
$$\sim \sqrt{\frac{\pi}{2}}\left[\langle a\Sigma^{-1}, a\rangle\left(\nu + \langle x\Sigma^{-1}, x\rangle\right)\right]^{-\frac{1}{4}}$$
$$\times \exp\left\{-\left[\langle a\Sigma^{-1}, a\rangle\left(\nu + \langle x\Sigma^{-1}, x\rangle\right)\right]^{\frac{1}{2}}\right\},$$

from (1.2) we find that, as $|x| \to \infty$,

$$f_{\nu,\Sigma,a}(x) \sim \frac{(\frac{\nu}{2})^{\frac{\nu}{2}}\left(\langle a\Sigma^{-1}, a\rangle\right)^{\frac{\nu+d-1}{4}}}{\Gamma(\frac{\nu}{2})(2\pi)^{\frac{d-1}{2}}\sqrt{|\Sigma|}} \frac{\exp\left\{\langle x\Sigma^{-1}, a\rangle\right\}}{\left(\nu + \langle x\Sigma^{-1}, x\rangle\right)^{\frac{\nu+d+1}{4}}}$$
$$\times \exp\left\{-\left[\langle a\Sigma^{-1}, a\rangle\left(\nu + \langle x\Sigma^{-1}, x\rangle\right)\right]^{\frac{1}{2}}\right\}$$
$$\sim c_{\nu,\Sigma,a}\left(\langle x\Sigma^{-1}, x\rangle\right)^{-\frac{\nu+d+1}{4}}$$
$$\times \exp\left\{-\left[\langle a\Sigma^{-1}, a\rangle\langle x\Sigma^{-1}, x\rangle\right]^{\frac{1}{2}} + \langle x\Sigma^{-1}, a\rangle\right\}.$$

□

Corollary 2.4 *Let d=1.*

(i) If $a > 0$, $x \to \infty$, then

$$f_{\nu,\sigma^2,a}(x) \sim \frac{1}{\sigma\Gamma(\frac{\nu}{2})}\left(\frac{\nu a}{2\sigma}\right)^{\frac{\nu}{2}} x^{-\frac{\nu}{2}-1}. \tag{2.8}$$

(ii) If $a > 0$, $x \to -\infty$, then

$$f_{\nu,\sigma^2,a}(x) \sim \frac{1}{\sigma\Gamma(\frac{\nu}{2})}\left(\frac{\nu a}{2\sigma}\right)^{\frac{\nu}{2}}|x|^{-\frac{\nu}{2}-1}\exp\left\{-\frac{2a|x|}{\sigma^2}\right\}. \tag{2.9}$$

(iii) If $a < 0$, $x \to \infty$, then

$$f_{\nu,\sigma^2,a}(x) \sim \frac{1}{\sigma\Gamma(\frac{\nu}{2})}\left(\frac{\nu|a|}{2\sigma}\right)^{\frac{\nu}{2}}x^{-\frac{\nu}{2}-1}\exp\left\{-\frac{2|a|x}{\sigma^2}\right\}. \tag{2.10}$$

(iv) If $a < 0$, $x \to -\infty$, then

$$f_{\nu,\sigma^2,a}(x) \sim \frac{1}{\sigma\Gamma(\frac{\nu}{2})}\left(\frac{\nu|a|}{2\sigma}\right)^{\frac{\nu}{2}}|x|^{-\frac{\nu}{2}-1}. \tag{2.11}$$

2.2 Asymptotic Distributions for Extremal and Record Values

Let now $d = 1$ and $\{X_n, n \geq 1\}$ a sequence of i.i.d. random variables with common Student's t-distribution function and let $M_n = \max_{1\leq j\leq n} X_j$.

Proposition 2.5 *(i) If pdf of $\mathcal{L}(X_1)$ is f_{ν,σ^2}, then, as $n \to \infty$,*

$$\mathcal{L}\left((K_1 n)^{-\frac{1}{\nu}}M_n\right) \Rightarrow \Phi_\nu,$$

where "$\Rightarrow$" means weak convergence of probability laws, Φ_ν is the Fréchet distribution

$$\Phi_\nu(x) = \begin{cases} \exp\{-x^{-\nu}\}, & \text{if } x > 0 \\ 0, & \text{if } x \leq 0, \end{cases}$$

and

$$K_1 = \frac{\Gamma(\frac{\nu+1}{2})\sigma^\nu}{\nu\sqrt{\pi}\Gamma(\frac{\nu}{2})}.$$

(ii) If pdf of $\mathcal{L}(X_1)$ is $f_{\nu,\sigma^2,a}$, $a > 0$, then, as $n \to \infty$,

$$\mathcal{L}\left((K_2 n)^{-\frac{2}{\nu}}M_n\right) \Rightarrow \Phi_{\frac{\nu}{2}},$$

where

$$K_2 = \frac{2(\frac{\nu a}{2\sigma})^{\frac{\nu}{2}}}{\nu\sigma\Gamma(\frac{\nu}{2})}.$$

(iii) If pdf of $\mathscr{L}(X_1)$ is f_{ν,a,σ^2}, $a < 0$, then, as $n \to \infty$,

$$\mathscr{L}\left(\frac{2|a|}{\sigma^2}M_n - \ln n - \left(\frac{\nu}{2}+1\right)\ln\ln n + \ln K_3\right) \Rightarrow \Lambda,$$

where Λ is the Gumbel distribution

$$\Lambda(x) = e^{-e^{-x}}, \quad x \in R^1,$$

and

$$K_3 = \frac{\nu^{\frac{\nu}{2}}\sigma^{\frac{\nu}{2}+3}}{2^{\nu+2}\Gamma(\frac{\nu}{2})}.$$

Proof (i) From Proposition 2.3 (i) with $d = 1$ and the l'Hospital's rule we have, as $x \to \infty$,

$$\int_x^\infty f_{\nu,\sigma^2}(u)\mathrm{d}u \sim \frac{c_{\nu,\sigma}}{\nu\sigma}\left(\frac{x}{\sigma}\right)^{-\nu} = K_1 x^{-\nu}, \tag{2.12}$$

where

$$c_{\nu,\sigma} = \frac{\Gamma\left(\frac{\nu+1}{2}\right)}{\sqrt{\pi}\Gamma\left(\frac{\nu}{2}\right)\sigma}.$$

The statement (i) is standard for Pareto-like distributions (see, e.g., [1, 2]).

(ii) From Corollary 2.4 (i) and the l'Hospital's rule we have that, as $x \to \infty$,

$$\int_x^\infty f_{\nu,\sigma^2,a}(u)\mathrm{d}u \sim K_2 x^{-\frac{\nu}{2}} \tag{2.13}$$

and the conclusion is analogs to (i).

(iii) From Corollary 2.4 (iii) and the l'Hospital's rule we find that, as $x \to \infty$,

$$\int_x^\infty f_{\nu,\sigma^2,a}(u)\mathrm{d}u \sim \frac{\sigma}{2|a|\Gamma(\frac{\nu}{2})}\left(\frac{\nu|a|}{2\sigma}\right)^{\frac{\nu}{2}} x^{-\frac{\nu}{2}-1}\exp\left\{-\frac{2|a|x}{\sigma^2}\right\}. \tag{2.14}$$

The statement (iii) is standard for gamma-like distributions (see, e.g., [1, 2]).

□

Now let us recall main results on limit theorems for record values in the sequences of i.i.d. random variables $\{X_n, n \geq 1\}$ with a common continuous distribution function F which will be applied to the case of Student's t-distributions.

The record times are $L_1 = 1$, $L_{n+1} = \min\{k : k > n, X_k > X_{L_n}\}$ for $n = 1, 2, \ldots$, and the record values are $R_n = X_{L_n}$, $n = 1, 2, \ldots$. Let $W(x) = -\log(1 - F(x))$ be the integrated hazard function and the associate distribution function $A(x) = 1 - e^{-\sqrt{W(x)}}$, $x \in R^1$. Let $l_{a,b}(x) = ax + b$, $a > 0$, $b \in R^1$, be a group of affine homeomorphisms of R^1 with the composition law

$$l_{a_1,b_1} * l_{a_2,b_2} = l_{a_1a_2,a_1b_2+b_1},$$

the unit element $l_{1,0}$ and the inverse $l_{a,b}^{-1} = l_{a^{-1},a^{-1}b}$.

The domain of attraction problem for record values using linear normalization was solved by Resnick (see [3] also [4]). It was proved that the class of all possible non-degenerated weak limit laws Q such that for suitable constants $a_n > 0$, $b_n \in R^1$, as $n \to \infty$,

$$\mathscr{L}\left(l_{a_n,b_n}^{-1}(R_n)\right) \Rightarrow Q$$

coincide with the class of laws $\Phi\left(-\log(-\log G(\cdot))\right)$, where Φ is a standard normal distribution and G is a l-max stable law, i. e. a non-degenerated distribution on R^1 such that for any $n \geq 2$ there exist constants $a_n > 0$, $b_n \in R^1$ satisfying

$$G^n(x) = G\left(l_{a_n,b_n}(x)\right), \quad x \in R^1.$$

As in the classical extreme value theory this class can be factorized into three linear types, saying that probability distributions F_1 and F_2 are of the same linear type it there exist constants $a > 0$, $b \in R^1$ such that

$$F_1(x) = F_2\left(l_{a,b}(x)\right), \quad x \in R^1.$$

In the classical case these types are generated by the Fréchet distribution Φ_γ, the Gumbel distribution Λ and the Weibull distribution

$$\Psi_\gamma(x) = \begin{cases} 1, & \text{if } x \geq 0, \\ \exp\{-(-x)^\gamma\}, & \text{if } x < 0, \quad \gamma > 0, \end{cases}$$

which correspond to generators of three types of the limiting laws for $\mathscr{L}\left(l_{a_n,b_n}(R_n)\right)$:

$$\tilde{\Phi}_\gamma(x) = \begin{cases} 0, & \text{if } x \leq 0, \\ \Phi(\log x^\gamma), & \text{if } x > 0, \quad \gamma > 0, \end{cases}$$

$$\tilde{\Psi}_\gamma(x) = \begin{cases} \Phi(\log(-x)^\gamma), & \text{if} \quad x < 0, \\ 1, \quad \text{if} \quad x \geq 0, & \gamma > 0, \end{cases}$$

and the standard normal distribution $\Phi(x)$, $x \in R^1$.

We say that F belongs to the record domain of attraction under linear normalization of the non-degenerated distribution Q ($F \in \mathrm{RDA}_l(Q)$ for short) if there exist constants $a_n > 0$ and $b_n \in R^1$ such that $\mathcal{L}(l^{-1}_{a_n,b_n}(R_n)) \Rightarrow Q$, as $n \to \infty$.

Duality theorem of Resnick says that $F \in \mathrm{RDA}_l(\tilde{\Phi}_\gamma) \Leftrightarrow A \in \mathrm{MDA}_l(\Phi_{\frac{\gamma}{2}})$, $F \in \mathrm{RDA}_l(\tilde{\Psi}_\gamma) \Leftrightarrow A \in \mathrm{MDA}_l(\Psi_{\frac{\gamma}{2}})$ and $F \in \mathrm{RDA}_l(\Phi) \Leftrightarrow A \in \mathrm{MDA}_l(\Lambda)$, where $\mathrm{MDA}_l(Q)$ denotes the maximum domain of attraction under linear normalization of the non-degenerated distribution Q (see, e.g., [3]). As a corollary we find that in the case of heavy-tailed distributions F the record values cannot have non-degenerate limiting distributions if we use linear normalization. Indeed, for the Pareto-like distributions F, satisfying, as $x \to \infty$,

$$1 - F(x) \sim Kx^{-\delta}, \quad \delta > 0,$$

the associate distributions A satisfy, as $x \to \infty$,

$$1 - A(x) \sim e^{-\sqrt{\delta \log x}}.$$

In this case $A \bar{\in} \mathrm{MDA}_l(\Phi_{\frac{\gamma}{2}}) \cup \mathrm{MDA}_l(\Psi_{\frac{\gamma}{2}}) \cup \mathrm{MDA}_l(\Lambda)$. This fact is an argument to consider limit theorems for the record values using power normalization.

Let

$$p_{\alpha,\beta}(x) = \alpha |x|^\beta \mathrm{sign} x, \quad \alpha > 0, \quad \beta > 0, \quad x \in R^1.$$

Observe that this class of functions form a group of homeomorphisms of R^1 with the composition law

$$p_{\alpha_1,\beta_1} * p_{\alpha_2,\beta_2} = p_{\alpha_1 \alpha_2^{\beta_1}, \beta_1 \beta_2},$$

the unit element $p_{1,1}$ and the inverse

$$p^{-1}_{\alpha,\beta} = p_{\alpha^{-\beta^{-1}}, \beta^{-1}}.$$

We say that F belongs to the record domain of attraction under power normalization of the non-degenerate distribution Q ($F \in \mathrm{RDA}_p(Q)$ for short) if there exist constants $\alpha_n > 0$, $\beta_n > 0$ such that, as $n \to \infty$, $\mathcal{L}\left(p^{-1}_{\alpha_n,\beta_n}(R_n)\right) \Rightarrow Q$.

A non-degenerate distribution function $\tilde{G}$ on R^1 is called p-max stable if for any $n \geq 2$ there exist constants $\tilde{\alpha}_n > 0$, $\tilde{\beta}_n > 0$ such that

$$\tilde{G}^n(x) = \tilde{G}(p_{\tilde{\alpha}_n,\tilde{\beta}_n}(x)), \quad x \in R^1.$$

Probability distributions F_1 and F_2 are of the same power type if there exist constants $\alpha > 0$, $\beta > 0$ such that $F_1(x) = F_2(p_{\alpha,\beta}(x))$, $x \in R^1$.

The class of non-degenerated limiting distributions for $\mathscr{L}(p^{-1}_{\alpha_n,\beta_n}(R_n))$, as $n \to \infty$, is equal to the class of law $\hat{\Phi}(-\log(-\log \hat{G}(\cdot)))$, where $\tilde{G}$ is a p-max stable law K, and is factorized to the six power types, generated by the distribution functions (see [5, 6]):

$$\hat{\Phi}_{1,\gamma}(x) = \begin{cases} 0, & \text{if } x \leq 1, \\ \Phi(\gamma \log\log x), & \text{if } x > 1, \quad \gamma > 0, \end{cases}$$

$$\hat{\Phi}_{2,\gamma}(x) = \begin{cases} 0, & \text{if } x \leq 0, \\ \Phi(-\gamma \log|\log x|), & \text{if } 0 < x < 1, \\ 1, & \text{if } x \geq 1, \quad \gamma > 0, \end{cases}$$

$$\hat{\Phi}_{3,\gamma}(x) = \begin{cases} 0, & \text{if } x \leq -1, \\ \Phi(-\gamma \log|\log|x||), & \text{if } -1 < x < 0, \\ 1, & \text{if } x \geq 0, \quad \gamma > 0, \end{cases}$$

$$\hat{\Phi}_{4,\gamma}(x) = \begin{cases} \Phi(-\gamma \log\log|x|), & \text{if } x < -1, \\ 1, & \text{if } x \geq -1, \quad \gamma > 0, \end{cases}$$

$$\hat{\Phi}_{5}(x) = \begin{cases} 0, & \text{if } x \leq 0, \\ \Phi(\log x), & \text{if } x > 0, \end{cases}$$

and

$$\hat{\Phi}_{6}(x) = \begin{cases} \Phi(-\log|x|), & \text{if } x < 0, \\ 1, & \text{if } x \geq 0. \end{cases}$$

There are the valid analog of Resnick's duality theorem and the principle of equivalent tails, which says that if continuous distribution functions F_1 and F_2 are such that $r(F_1) = r(F_2)$ and $1 - F_1(x) \sim 1 - F_2(x)$, as $x \uparrow r(F_1)$, then $F_1 \in \mathrm{RDA}_p(Q)$ if and only if $F_2 \in \mathrm{RDA}_p(Q)$ with the same normalizing constants, where $r(F) = \sup\{x : F(x) < 1\}$ and Q is a non-degenerate limiting distribution for record values using power normalization.

The following analog of classical R. von Mises theorem [7] holds true.

Theorem 2.6 [8]. *Assume that the integrated hazard function $W(x)$ is differentiable in some neighborhood of $r(F)$. Then:*

(i) if $r(F) = \infty$ and

$$\lim_{x \to \infty} \frac{W'(x) x \log x}{\sqrt{W(x)}} = \gamma, \quad \gamma > 0,$$

then $F \in RDA_p(\hat{\Phi}_{1,\gamma})$;

(ii) *if* $0 < r(F) < \infty$ *and*

$$\lim_{x \uparrow r(F)} \frac{W'(x)x \log\left(\frac{r(F)}{x}\right)}{\sqrt{W(x)}} = \gamma, \quad \gamma > 0,$$

then $F \in RDA_p(\hat{\Phi}_{2,\gamma})$*;*

(iii) *if* $r(F) = 0$ *and*

$$\lim_{x \uparrow 0} \frac{W'(x)x \log|x|}{\sqrt{W(x)}} = \gamma, \quad \gamma > 0,$$

then $F \in RDA_p(\hat{\Phi}_{3,\gamma})$*;*

(iv) *if* $r(F) < 0$ *and*

$$\lim_{x \uparrow r(F)} \frac{W'(x)|x| \log\left(\frac{x}{r(F)}\right)}{\sqrt{W(x)}} = \gamma, \quad \gamma > 0,$$

then $F \in RDA_p(\hat{\Phi}_{4,\gamma})$*;*

(v) *if* W *is twice differentiable in some neighborhood of* $r(F)$ *and*

$$\lim_{x \uparrow r(F)} W(x)\left(\frac{W''(x)}{(W'(x))^2} + \frac{1}{xW'(x)}\right) = 0, \tag{2.15}$$

then for $0 < r(F) \leq \infty$ $F \in RDA_p(\hat{\Phi}_5)$ *and for* $r(F) \leq 0$ $F \in RDA_p(\hat{\Phi}_6)$.

Proposition 2.7

(i) *If pdf of* F *is* f_{ν,σ^2}, *then* $F \in RDA_p(\hat{\Phi}_5)$.

(ii) *If pdf of* F *is* $f_{\nu,\sigma^2,a}$, $a > 0$, *then* $F \in RDA_p(\hat{\Phi}_5)$.

(iii) *If pdf of* F *is* $f_{\nu,\sigma^2,a}$, $a < 0$, *then* $F \in RDA_l(\Phi)$.

Proof

(i) From the principle of equivalent tails and (2.12) it is enough to check (2.15) with $r(F) = \infty$ and the integrated hazard function

$$W(x) = \nu \ln x - \ln K_1.$$

Indeed,

$$\frac{W''(x)}{(W'(x))^2} + \frac{1}{xW'(x)} = \frac{-\frac{\nu}{x^2}}{\left(\frac{\nu}{x}\right)^2} + \frac{1}{\nu} \equiv 0.$$

(ii) From the principle of equivalent tails and (2.13) it is enough to check (2.15) with $r(F) = \infty$ and the integrated hazard function

$$W(x) = \frac{\nu}{2} \ln x - \ln K_2.$$

Again we find that

$$\frac{W''(x)}{(W'(x))^2} + \frac{1}{xW'(x)} = \frac{-\frac{\nu}{2x^2}}{\left(\frac{\nu}{2x}\right)^2} + \frac{2}{\nu} \equiv 0.$$

(iii) From (2.14) and the principle of equivalent tails it is enough to consider the integrated hazard function

$$W(x) = \left(\frac{\nu}{2} + 1\right) \ln x + \frac{2|a|}{\sigma^2} x - \ln K_3,$$

where

$$K_3 = \frac{\sigma}{2|a|\Gamma\left(\frac{\nu}{2}\right)} \left(\frac{\nu|a|}{2\sigma}\right)^{\frac{\nu}{2}}.$$

The corresponding associated distribution

$$\begin{aligned} 1 - A(x) &= \exp\left\{-\sqrt{\left(\frac{\nu}{2} + 1\right) \ln x + \frac{2|a|}{\sigma^2} x - \ln K_3}\right\} \\ &\sim \exp\left\{-\sqrt{\frac{2|a|}{\sigma^2} x}\right\}, \quad \text{as} \quad x \to \infty. \end{aligned}$$

Using again the principle of equivalent tails, Resnick's duality theorem and criteria from the classical extreme value theory we easily find that $A \in \mathrm{MDA}_l(\Lambda)$ and thus $F \in \mathrm{RDA}_l(\Phi)$. □

References

1. Embrechts, P., Klüppelberg, C., Mikosch, T.: Modelling Extremal Events for Insurance and Finance. Springer, Berlin (1997)
2. Leadbetter, M.R., Lindgren, G., Rootzen, H.: Extremes and Related Properties of Random Sequences and Processes. Springer, Berlin (1983)
3. Resnik, S.I.: Limit laws for record values. Stoch. Processes Appl. **1**, 67–82 (1973)
4. Tata, M.N.: On outstanding values in a sequence of random variables. Z. Wahrscheinlichkeitstheor. vewr. Geb. **12**(1), 9–20 (1969)

5. Mohan, N.R., Ravi, S.: Max domains of attraction of univariate and multivariate p-max stable laws. Teor. Veroyatnost. i Primenen, **37**(4), 709–721 (1992)
6. Pantcheva, E.: Limit theorems for extreme order statistics under nonlinear normalization. In: Lecture Notes in Math., vol. 1155, pp. 284–309. Springer, Berlin, (1985)
7. von Mises, R.: La distribution de la plus grande de n valeurs. Revue Mathématique de l'Union Interbalkanique (Athens) **1**, 141–160 (1936)
8. Grigelionis, B.: Limit theorems for record values using power normalization. Lith. Math. J. **46**(4), 398–405 (2006)

Chapter 3
Preliminaries of Lévy Processes

3.1 Lévy-Itô Decomposition

Let $(\Omega, \mathscr{F}, \mathrm{P})$ be a probability space and $(R^d, \mathscr{B}(R^d), \langle\cdot,\cdot\rangle)$ be a d-dimensional Euclidean space R^d with the σ-algebra of Borel subsets $\mathscr{B}(R^d)$, the scalar product $\langle x, y\rangle = \sum_{j=1}^d x_j y_j$ for row vectors $x = (x_1, \dots, x_d)$, $y = (y_1, \dots, y_d)$, and the norm $|x| = \sqrt{\langle x, x\rangle}$.

We are assuming that the reader is familiar with the foundations of probability theory based on the measure theory.

A mapping $X : R_+ \times \Omega \to R^d$ such that for each $B \in \mathscr{B}(R^d)$ and $t \geq 0$ $\{\omega : X(t, \omega) \in B\} \in \mathscr{F}$ is called a d-dimensional stochastic process.

For fixed $\omega \in \Omega$ a function $X(\cdot, \omega)$ is called a sample path of X. Later we shall use the notation $X = \{X_t, t \geq 0\}$.

If for each $t \geq 0$ and $\varepsilon > 0$

$$\lim_{h\to 0} \mathrm{P}\{|X_{t+h} - X_t| > \varepsilon\} = 0,$$

a process $X = \{X_t, t \geq 0\}$ is called stochastically continuous.

Definition 3.1 A d-dimensional stochastic process $X = \{X_t, t \geq 0\}$ is an additive process if the following conditions are satisfied:

(1) for any $n \geq 1$ and $0 \leq t_0 < t_1 < \cdots < t_n$, increments $X_{t_0}, X_{t_1} - X_{t_0}, \dots, X_{t_n} - X_{t_{n-1}}$ are independent;
(2) $X_0 = 0$ P-a.e.;
(3) X is stochastically continuous;
(4) P-a.e. sample paths are right-continuous in $t \geq 0$ and have left limits in $t > 0$.

An additive process in law is a stochastic process satisfying (1)–(3).

Let $X = \{X_t, t \geq 0\}$ be a d-dimensional additive process.

Let $U_\varepsilon = \{x \in R^d : |x| > \varepsilon\}$, $\varepsilon > 0$, $\mathscr{B}_\varepsilon(R^d) = \mathscr{B}(R^d) \cap U_\varepsilon$.

B. Grigelionis, *Student's* t-*Distribution and Related Stochastic Processes*, SpringerBriefs in Statistics, DOI: 10.1007/978-3-642-31146-8_3,

For $B \in \mathscr{B}_\varepsilon(R^d)$ define

$$p(t, B) = \sum_{0 \le s \le t} 1_B(X_s - X_{s-})$$

and

$$X_t^B = \sum_{0 \le s \le t} (X_s - X_{s-}) 1_B(X_s - X_{s-}), \quad t \ge 0.$$

The following properties hold true (see, e.g., [1–3]).

(i) For each $B \in \mathscr{B}_\varepsilon(R^d)$, $t > 0$, the function

$$\mathrm{E} p(t, B) := \Pi(t, B) < \infty.$$

and is continuous in t.
The stochastic process $p(t, B)$, $t \ge 0$ is a Poisson additive process with mean function $\Pi(t, B)$, $t \ge 0$, i. e. it satisfies the assumptions (1)–(4) and for each $t > 0$, $k = 0, 1, \ldots$

$$\mathrm{P}\{p(t, B) = k\} = e^{-\Pi(t,B)} \frac{(\Pi(t, B))^k}{k!}.$$

Moreover,

$$\int_{R^d \setminus \{0\}} |x|^2 \wedge 1 \Pi(t, \mathrm{d}x) < \infty.$$

(ii) For each $B_1, \ldots, B_m \in \mathscr{B}_\varepsilon(R^d)$ such that $B_j \cap B_k = \emptyset$, $j \ne k$, stochastic processes

$$\left\{X_t^{B_1}, t \ge 0\right\}, \ldots, \left\{X_t^{B_m}, t \ge 0\right\} \quad \text{and} \quad \left\{X_t - \sum_{j=1}^m X_j^{B_j}, t \ge 0\right\}$$

are additive mutually independent processes and for each $\varepsilon > 0$, $t > 0$

$$\mathrm{E}|X_t - X_t^{U_\varepsilon}|^2 < \infty.$$

(iii) Let $0 < \varepsilon_n \downarrow 0$, as $n \to \infty$, and $\Delta_k = \left\{x \in R^d : \varepsilon_k < |x| \le \varepsilon_{k-1}\right\}$, $k = 2, 3, \ldots$, $\Delta_1 = \left\{x \in R^d : |x| > \varepsilon_1\right\}$.
There exists a subsequence $\{n_k, k = 1, 2, \ldots\}$ such that, as $k \to \infty$, the sequence

$$X_t^{(k)} := X_t - X_t^{\Delta_1} - \sum_{j=2}^{n_k} (X_t^{\Delta_j} - EX_t^{\Delta_j}), \quad t \geq 0,$$

converges uniformly on each finite time interval P-a.e. to the continuous Gaussian additive process $X^0 = \{X_t^0, t \geq 0\}$ such that

$$\mathrm{E} e^{i\langle z, X_t^0 \rangle} = \exp\left\{ i\langle z, a(t)\rangle - \frac{1}{2}\langle zA(t), z\rangle \right\}, \quad z \in R^d,$$

where $a(t)$, $t \geq 0$, is a continuous d-dimensional function and $A(t)$ is a continuous symmetric nonnegative definite $d \times d$ matrix valued function.

(iv) For each $z \in R^d$ and $t > 0$

$$\mathrm{E} \exp\{i\langle z, X_t\rangle\} = \exp\left\{ i\langle z, a(t)\rangle - \frac{1}{2}\langle zA(t), z\rangle + \int\limits_{R^d \setminus \{0\}} \left(e^{i\langle z, x\rangle} - 1 - i\langle z, x\rangle 1_{\{|x| \leq 1\}} \right) \Pi(t, \mathrm{d}x) \right\}, \quad (3.1)$$

implying the Lévy-Khinchine formula as $t = 1$.

Definition 3.2 A d-dimensional additive process $X = \{X_t, t \geq 0\}$ is called a Lévy process if it is temporally homogeneous, i.e., for each $s, t > 0$,

$$\mathscr{L}(X_{t+s} - X_s) = \mathscr{L}(X_t).$$

Definition 3.3 A d-dimensional stochastic process $X = \{X_t, t \geq 0\}$ is called a Lévy process in law if it is temporally homogeneous and satisfies the assumptions (1)–(3).

An additive process is a Lévy one if and only if the functions $a(t)$, $A(t)$ and $\Pi(t, B)$, $t \geq 0$, are linear in t, i.e., $a(t) = at$, $A(t) = At$ and $\Pi(t, B) = \Pi(B)t$. The triplet (a, A, Π), where $a \in R^d$, A is a symmetric nonnegative definite $d \times d$ matrix and $\Pi(B)$, $B \in \mathscr{B}(R_0^d)$, is a measure such that

$$\int\limits_{R_0^d} |x|^2 \wedge 1 \Pi(\mathrm{d}x) < \infty,$$

is called the triplet of Lévy characteristics; $R_0^d := R^d \setminus \{0\}$. This triplet uniquely defines the finite dimensional distributions $\mathscr{L}(X_{t_1}, X_{t_2}, \ldots, X_{t_n})$, $0 \leq t_1 < t_2 < \cdots < t_n, n \geq 1$.

The class of Lévy triplets corresponds one-to-one with the class of Lévy processes in law.

A d-dimensional Lévy process X with the triplet of Lévy characteristics $(0, I_d, 0)$, where I_d is the $d \times d$ unit matrix, is called the standard d-dimensional Brownian motion .

Definition 3.4 A probability distribution μ on R^d is called infinitely divisible if, for any positive integer n, there exists a probability measure μ_n on R^d such that $\mu = \underbrace{\mu_n * \cdots * \mu_n}_{n \text{ times}}$. Here "$*$" means the convolution of probability distributions.

We shall write that $\mu \in ID(R^d)$.

From the celebrated Lévy-Khinchine formula and (3.1) it follows that the class of infinitely distributions μ corresponds one-to-one with the class of Lévy processes in law by means of the equality

$$\begin{aligned} &\mathrm{E}\exp\{i\langle z, X_1\rangle\} \\ &\quad = \exp\left\{ i\langle z, a\rangle - \frac{1}{2}\langle zA, z\rangle + \int\limits_{R_0^d} \left(e^{i\langle z,x\rangle} - 1 - i\langle z, x\rangle 1_{\{|x|\le 1\}}\right) \Pi(\mathrm{d}x) \right\} \\ &\quad = \int\limits_{R^d} e^{i\langle z,x\rangle}\mu(\mathrm{d}x). \end{aligned}$$

For each Lévy process in law $X = \{X_t, t \ge 0\}$ there exists a modification $Y = \{Y_t, t \ge 0\}$ with right-continuous sample paths in $t \ge 0$, having left limits in $t > 0$ and satisfying $P(X_t \ne Y_t) = 0, t \ge 0$.

3.2 Self-Decomposable Lévy Processes

Definition 3.5 A probability distribution μ on R^d is called self-decomposable, or of class $L(R^d)$, if, for any $b > 1$, there exists a probability measure ρ_b on R^d such that

$$\hat{\mu}(z) = \hat{\mu}(b^{-1}z)\hat{\rho}_b(z), \quad z \in R^d, \tag{3.2}$$

where $\hat{\mu}$ means the characteristic function of the probability distribution μ on R^d.

If μ is self-decomposable, then μ is infinitely divisible and, for any $b > 1$, ρ_b in the decomposition (3.2) is uniquely determined and ρ_b is infinitely divisible.

Definition 3.6 A Lévy process $X = \{X_t, t \ge 0\}$ in law is said to be self-decomposable if the probability distribution $\mathscr{L}(X_1)$ is self-decomposable.

The Gaussian Lévy processes in law are, obviously, self-decomposable, because in this case (3.2) is satisfied with

$$\hat{\rho}_b(z) = \exp\left\{ i\langle z, (1-b^{-1})a\rangle - \frac{1}{2}\langle z(1-b^{-2})A, z\rangle \right\}.$$

A criterion of self-decomposability of non-Gaussian $\mu \in ID(R^d)$ with the triplet (a, A, Π) of Lévy characteristics will be formulated using the canonical polar decomposition of a Lévy measure Π (see Remark 16 in [4], Lemma 1 in [5] and Proposition 2 in [6]).

Write

$$S^{d-1} = \left\{ x \in R^d : |x| = 1 \right\}, \quad K = \int\limits_{R_0^d} |x|^2 \wedge 1 \Pi(\mathrm{d}x) > 0.$$

Proposition 3.7 *There exists a pair (λ, Π_ξ), where λ is a probability measure on S^{d-1} and Π_ξ is a σ-finite measure on $(0, \infty)$ such that $\Pi_\xi(C)$ is measurable in $\xi \in S^{d-1}$ for every $C \in \mathscr{B}((0, \infty))$,*

$$\int\limits_0^\infty r^2 \wedge 1 \Pi_\xi(dr) \equiv K \tag{3.3}$$

and

$$\Pi(B) = \int\limits_{S^{d-1}} \lambda(d\xi) \int\limits_0^\infty 1_B(r\xi) \Pi_\xi(dr), \quad B \in \mathscr{B}(B_0^d). \tag{3.4}$$

If a pair (λ', Π'_ξ) satisfies (3.3) and (3.4), then $\lambda' = \lambda$ and $\Pi_\xi = \Pi'_\xi$ λ-a.e.

Proof *Existence.* Consider the probability space $(R_0^d, \mathscr{B}(R_0^d), \mathrm{P}_\Pi)$, where

$$\mathrm{P}_\Pi(B) = K^{-1} \int\limits_B |x|^2 \wedge 1 \Pi(\mathrm{d}x), \quad B \in \mathscr{B}(R_0^d).$$

Let $N(x) = x$, $R(x) = |x|$, $\Xi(x) = \frac{x}{|x|}$, $x \in R_0^d$, $\lambda(B) = \mathrm{P}_\Pi\{\Xi \in B\}$, $B \in \mathscr{B}(S^{d-1})$, $\Pi_\xi^0(C) = \mathrm{P}_\Pi\{R \in C | \Xi = \xi\}$ (a regular version of the conditional distribution), and

$$\Pi_\xi(C) = \int\limits_C K(r^2 \wedge 1)^{-1} \Pi_\xi^0(\mathrm{d}r), \quad C \in \mathscr{B}((0, \infty)), \quad \xi \in S^{d-1}.$$

The pair (λ, Π_ξ) satisfies (3.3) and (3.4). Indeed,

$$\int_0^\infty (r^2 \wedge 1)\Pi_\xi(\mathrm{d}r) = K \int_0^\infty \Pi_\xi^0(\mathrm{d}r) \equiv K,$$

and for every nonnegative measurable function $f(x)$, $x \in R_0^d$,

$$\begin{aligned}\int_{S^{d-1}} \lambda(\mathrm{d}\xi) \int_0^\infty f(r\xi)\Pi_\xi(\mathrm{d}r) &= \int_{S^{d-1}} \lambda(\mathrm{d}\xi) \int_0^\infty \frac{Kf(r\xi)}{r^2 \wedge 1}\Pi_\xi^0(\mathrm{d}r)\\ &= \mathrm{E}_\Pi\left[\mathrm{E}_\Pi\left(\frac{Kf(R\Xi)}{R^2 \wedge 1}|\Xi\right)\right]\\ &= \mathrm{E}_\Pi\left(\frac{Kf(N)}{R^2 \wedge 1}\right) = \int_{R_0^d} f(x)\Pi(\mathrm{d}x).\end{aligned}$$

It remains to take $f(x) = 1_B(x)$, $B \in \mathscr{B}(R_0^d)$.
Uniqueness. Let

$$\Pi(B) = \int_{S^{d-1}} \lambda'(\mathrm{d}\xi) \int_0^\infty 1_B(r\xi)\Pi'_\xi(\mathrm{d}r) \tag{3.5}$$

and

$$\int_0^\infty (r^2 \wedge 1)\Pi'_\xi(\mathrm{d}r) \equiv K. \tag{3.6}$$

Then, for all $B \in \mathscr{B}(S^{d-1})$, from (3.3)–(3.6) we find that

$$\begin{aligned}&\int_{R_0^d} 1_B\left(\frac{x}{|x|}\right) K^{-1}(|x|^2 \wedge 1)\Pi(\mathrm{d}x)\\ &= \int_{S^{d-1}} \lambda(\mathrm{d}\xi) \int_0^\infty 1_B(\xi)K^{-1}(r^2 \wedge 1)\Pi_\xi(\mathrm{d}r) = \lambda(B)\end{aligned}$$

and

$$\int_{R_0^d} 1_B\left(\frac{x}{|x|}\right) K^{-1}(|x|^2 \wedge 1)\Pi(\mathrm{d}x)$$

$$= \int_{S^{d-1}} \lambda'(d\xi) \int_0^\infty 1_B(\xi) K^{-1}(r^2 \wedge 1)\Pi'_\xi(dr) = \lambda'(B),$$

proving that $\lambda = \lambda'$.

Finally, for every nonnegative measurable function $h(r)$, $r > 0$,

$$\int_{R_0^d} h(|x|)\Pi(dx) = \int_{S^{d-1}} \lambda(d\xi) \int_0^\infty h(r)\Pi_\xi(dr) = \int_{S^{d-1}} \lambda(d\xi) \int_0^\infty h(r)\Pi'_\xi(dr),$$

implying that $\Pi_\xi = \Pi'_\xi$ λ-a.e. □

Proposition 3.8 *[7]. If*

$$\Pi(B) = \int_B g(x)dx, \quad B \in \mathscr{B}(R_0^d), \tag{3.7}$$

then (3.3) *and* (3.4) *hold with*

$$\lambda(d\xi) = c(\xi)d\xi,$$
$$\Pi_\xi(dr) = r^{d-1} g(r\xi)c^{-1}(\xi),$$

where

$$c(\xi) = K^{-1} \int_0^\infty (r^2 \wedge 1) r^{d-1} g(r\xi)dr,$$

assuming that

$$K := \int_{R_0^d} (|x|^2 \wedge 1) g(x)dx > 0.$$

Proof Write

$$\begin{aligned}
x_1 &= r\cos\varphi_1,\\
x_2 &= r\sin\varphi_1\cos\varphi_2,\\
&\cdots\cdots\cdots\cdots\\
x_{d-1} &= r\sin\varphi_1\sin\varphi_2\cdots\sin\varphi_{d-1}\cos\varphi_{d-1},\\
x_d &= r\sin\varphi_1\sin\varphi 2\cdots\sin\varphi_{d-2}\sin\varphi_{d-1},
\end{aligned}$$

where $r \geq 0, 0 \leq \varphi_1 \leq \pi, \ldots, 0 \leq \varphi_{d-2} \leq \pi, 0 \leq \varphi_{d-1} < 2\pi$. It is well-known that the Jacobian

$$J = \frac{D(x_1, x_2, \ldots, x_d)}{D(r, \varphi_1, \varphi_2, \ldots, \varphi_{d-1})} = r^{d-1} \sin^{d-2} \varphi_1 \sin^{d-3} \varphi_2 \cdots \sin \varphi_{d-2}.$$

Denoting $\xi = \frac{x}{r}$ and $\mathrm{d}\xi = \sin^{d-2} \varphi_1 \sin^{d-3} \varphi_2 \cdots \sin \varphi_{d-2} \mathrm{d}\varphi_1 \mathrm{d}\varphi_2 \cdots \mathrm{d}\varphi_{d-2} \mathrm{d}\varphi_{d-1}$, for any Borel measurable and integrable with respect to the Lebesgue measure on R^d function $f(x)$, we find that

$$\int_{R^d} f(x)\mathrm{d}x = \int_{S^{d-1}} \mathrm{d}\xi \int_0^\infty f(r\xi) r^{d-1} \mathrm{d}r = \int_{S^{d-1}} c(\xi)\mathrm{d}\xi \int_0^\infty f(r\xi) r^{d-1} c^{-1}(\xi) \mathrm{d}r \tag{3.8}$$

and apply formula (3.8) to the functions $f_B(x) = g(x)1_B(x), x \in R^d, B \in \mathscr{B}(R_0^d)$. The identity (3.3) is trivially satisfied. □

The following criterion of self-decomposability of a probability distribution $\mu \in ID(R^d)$ with the triplet of Lévy characteristics (a, A, Π) is well-known (see Theorem 15.10 of [2] and [8]).

Theorem 3.9 *A probability distribution $\mu \in ID(R^d)$ or a Lévy process in law with the triplet (a, A, Π) of Lévy characteristics is self-decomposable if and only if in* (3.4)

$$\Pi_\xi(dr) = \frac{k_\xi(r)}{r} dr,$$

where a nonnegative function $k_\xi(r)$ is measurable in $\xi \in S^{d-1}$ and decreasing in $r > 0$ for λ-a.e. ξ.

Corollary 3.10 *If* (3.7) *is satisfied, then $\mu \in ID(R^d)$ or a corresponding Lévy process in law with the triplet (a, A, Π) of Lévy characteristics is self-decomposable if and only if the function $k_\xi(r) := r^d g(r\xi)$ is decreasing in $r > 0$ for a.e. $\xi \in S^{d-1}$ with respect to the Lebesgue surface measure on S^{d-1}.*

3.3 Lévy Subordinators

Definition 3.11 An univariate Lévy process with nonnegative increments is called a Lévy subordinator.

The class of Lévy subordinators correspond one-to-one with the class $ID(R_+)$ of infinitely divisible distributions on R_+. It is well-known (see, e.g., [2, 3, 9, 10]) that for $\tau \in ID(R_+)$ the Laplace exponent

$$\psi(\theta) := -\log\left(\int_0^\infty e^{-\theta u}\tau(du)\right) = \beta_0\theta + \int_0^\infty \left(1 - e^{-\tau u}\rho(du)\right), \quad \theta \geq 0,$$

is defined uniquely by the characteristics (β_0, ρ), where $\beta_0 \geq 0$ and ρ is a σ-finite measure on $(0, \infty)$, satisfying

$$\int_0^\infty (u \wedge 1)\rho(du) < \infty.$$

Extending the Thorin class and following Bondesson [11], we introduce the scale of Thorin classes $T_\varkappa(R_+), 0 < \varkappa \leq \infty$, as increasing subclasses of $ID(R_+)$ such that $T_\infty(R_+) = ID(R_+)$, where $T_\infty(R_+)$ is the minimal class of probability distributions on R_+, closed under convolutions and weak limits, containing all classes $T_\varkappa(R_+)$, $\varkappa > 0$.

Definition 3.12 An infinitely divisible distribution τ on R_+ with the characteristics (β_0, ρ) is of the Thorin class $T_\varkappa(R_+)$, $\varkappa > 0$, if $\rho(dt) = l(t)dt$ and $k_\varkappa(t) := t^{2-\varkappa}l(t)$, $t \geq 0$, is completely monotone, i.e., $k_\varkappa$ is infinitely differentiable and $(-1)^n k_\varkappa^{(n)}(t) \geq 0$ for all $n \geq 0$ and $t > 0$.

Lévy subordinators corresponding to the distributions from $T_\varkappa(R_+), 0 < \varkappa < \infty$, are called the Thorin's subordinators.

According to Bernstein's theorem (see, e.g., [12]) there exists a unique positive measure $Q_\varkappa$ on R_+ such that

$$k_\varkappa(t) = \int_0^\infty e^{-vt} Q_\varkappa(dv), \quad t > 0,$$

and

$$Q_\varkappa(\{0\}) = \lim_{t\to\infty} k_\varkappa(t).$$

Write

$$a_\varkappa(t) = t^{-\varkappa}\int_0^t v^{\varkappa-1}e^{-v}dv + t^{-\varkappa+1}\int_t^\infty v^{\varkappa-2}e^{-v}dv, \quad t > 0,$$

and observe that, as $t \to \infty$,

$$a_\varkappa(t) \sim \Gamma(\varkappa)t^{-\varkappa},$$

and

$$a_{\varkappa}(t) \sim \begin{cases} (\varkappa(1-\varkappa))^{-1}, & \text{if } 0 < \varkappa < 1, \\ \log \dfrac{1}{t}, & \text{if } \varkappa = 1, \\ \Gamma(\varkappa - 1)t^{1-\varkappa}, & \text{if } \varkappa > 1, \end{cases} \tag{3.9}$$

as $t \to 0$.

Proposition 3.13 *An infinitely divisible distribution τ on R_+ with the characteristics (β_0, ρ) is of the Thorin class $T_{\varkappa}(R_+)$, $\varkappa > 0$, if and only if the Laplace exponent*

$$\psi_{\varkappa}(\theta) = \begin{cases} \beta_0\theta + \Gamma(\varkappa - 1) \int\limits_0^{\infty} \left(v^{-\varkappa+1} - (\theta + v)^{-\varkappa+1}\right) Q_{\varkappa}(dv), & \text{if } \varkappa \neq 1, \\ \beta_0\theta + \int\limits_0^{\infty} \log\left(1 + \frac{\theta}{v}\right) Q_1(dv), & \text{if } \varkappa = 1, \end{cases} \tag{3.10}$$

where the measure $Q_{\varkappa}$, called the Thorin measure, satisfies

$$\int\limits_0^{\infty} a_{\varkappa}(v) Q_{\varkappa}(dv) < \infty, \tag{3.11}$$

implying that $Q_{\varkappa}(\{0\}) = 0$ for $\varkappa \geq 1$.

Proof We have that

$$\int\limits_0^{\infty} (t \wedge 1) l(t) \mathrm{d}t = \int\limits_0^{\infty} (t \wedge 1) t^{\varkappa-2} \int\limits_0^{\infty} e^{-tv} Q_{\varkappa}(\mathrm{d}v)\mathrm{d}t = \int\limits_0^{\infty} a_{\varkappa}(v) Q_{\varkappa}(\mathrm{d}v)$$

and

$$\begin{aligned} \psi_{\varkappa}(\theta) &= \beta_0\theta + \int\limits_0^{\infty} \left(1 - e^{-\theta t}\right) t^{\varkappa-2} \int\limits_0^{\infty} e^{-vt} Q_{\varkappa}(\mathrm{d}v)\mathrm{d}t \\ &= \beta_0\theta + \int\limits_0^{\infty} \left(\int\limits_0^{\infty} t^{\varkappa-2} \left(1 - e^{-\theta t}\right) e^{-vt} \mathrm{d}t \right) Q_{\varkappa}(\mathrm{d}v). \end{aligned}$$

However, for $0 < \varkappa < 1$,

$$\int\limits_0^{\infty} t^{\varkappa-2} \left(1 - e^{-\theta t}\right) e^{-vt} \mathrm{d}t = -\int\limits_0^{\infty} e^{-vt} t^{\varkappa-2} \sum_{k=1}^{\infty} \frac{(-t\theta)^k}{k!} \mathrm{d}t$$

$$= -\sum_{k=1}^{\infty} \frac{(-\theta)^k}{k!} v^{-k-\theta+1} \Gamma(k+\varkappa-1)$$

$$= -v^{-\varkappa+1} \sum_{k=1}^{\infty} \frac{\Gamma(k+\varkappa-1)}{k!} \left(-\frac{\theta}{v}\right)^k$$

$$= v^{-\varkappa+1} \Gamma(\varkappa-1) \left(1 - \left(1+\frac{\theta}{v}\right)^{-\varkappa+1}\right),$$

for $\varkappa = 1$, as a Froullani integral,

$$\int_0^{\infty} \left(1 - e^{-\theta t}\right) t^{-1} e^{-vt} \mathrm{d}t = \log\left(1+\frac{\theta}{v}\right)$$

and, for $\varkappa > 1$,

$$\int_0^{\infty} t^{\varkappa-2} \left(1 - e^{-\theta t}\right) e^{-vt} \mathrm{d}t = \frac{1}{\varkappa-1} \int_0^{\infty} \left(1 - e^{-\theta t}\right) e^{-vt} \mathrm{d}t^{\varkappa-1}$$

$$= \frac{1}{\varkappa-1} \int_0^{\infty} t^{\varkappa-1} \left[ve^{-vt} - (\theta+v)e^{-(\theta+v)t}\right] \mathrm{d}t$$

$$= \frac{\Gamma(\varkappa)}{\varkappa-1} \left(v^{-\varkappa+1} - (\theta+v)^{-\varkappa+1}\right)$$

$$= \Gamma(\varkappa-1) \left(v^{-\varkappa+1} - (\theta+v)^{-\varkappa+1}\right).$$

□

Remark 3.14 Having in mind (3.9), inequality (3.11) is satisfied if and only if the measure $Q_\varkappa$ is a Radon measure such that for $\varkappa \neq 1$

$$\int_0^1 u^{0\wedge(1-\varkappa)} Q_\varkappa(\mathrm{d}u) < \infty \quad \text{and} \quad \int_1^{\infty} u^{-\varkappa} Q_\varkappa(\mathrm{d}u) < \infty,$$

and for $\varkappa = 1$

$$\int_0^1 \log\left(\frac{1}{u}\right) Q_1(\mathrm{d}u) < \infty \quad \text{and} \quad \int_1^{\infty} u^{-1} Q_1(\mathrm{d}u) < \infty.$$

Recall now that the families of Tweedie or power-variance distributions

$$\left\{Tw_p(\alpha, \lambda), \alpha > 0, \lambda > 0\right\}, \quad p \in R^1 \backslash [0, 1),$$

are defined as exponential dispersion models (see [13–15]), satisfying the following properties: for each $\alpha > 0$, $\lambda > 0$ and given p

$$\int_{R^1} x Tw_p(\alpha, \lambda)(dx) = \alpha,$$

$$\int_{R^1} (x-\alpha)^2 Tw_p(\alpha, \lambda)(dx) = \lambda^{-1}\alpha^p$$

and $Tw_0(\alpha, \lambda) := N(\alpha, \lambda^{-1})$, $\alpha \in R^1$, $\lambda > 0$. It is known that for $p \geq 1$ $Tw_p(\alpha, \lambda) \in ID(R_+)$. Moreover, for $p > 1, \alpha > 0, \lambda > 0$ $Tw_p(\alpha, \lambda) \in T_{\frac{1}{p-1}}(R_+)$, because their characteristics are

$$\left(0, c_{p,\lambda} t^{-2+\frac{1}{p-1}} \exp\left\{-\frac{\alpha^{1-p}}{p-1}\lambda t\right\} dt\right),$$

where

$$c_{p,\lambda} = \frac{\lambda^{\frac{1}{p-1}}}{\Gamma\left(\frac{p}{p-1}\right)(p-1)^{\frac{p}{p-1}}},$$

and $Tw_1(\alpha, \lambda) \in ID(R_+)$ with characteristics $(0, \alpha\lambda\varepsilon_{\lambda^{-1}}(dt))$. The Thorin measure $Q_{\frac{1}{p-1}}$ of $Tw_p(\alpha, \lambda)$, $p > 1$, obviously, equals

$$c_{p,\lambda}\varepsilon_{\frac{\alpha^{1-p}}{p-1}}(dt).$$

Theorem 3.15 [16].

(i) Thorin classes $T_\varkappa(R_+)$, $0 < \varkappa < \infty$, are increasing, closed under convolutions and weak limits;
(ii) $T_\infty(R_+) = ID(R_+)$;
(iii) Thorin classes $T_\varkappa(R_+)$, $0 < \varkappa \leq \infty$, are generalized convolutions of Tweedie distributions $T_{\frac{\varkappa+1}{\varkappa}}(\alpha, \lambda)$, $\alpha > 0$, $\lambda > 0$.

Proof Because for $0 < \varkappa_1 < \varkappa_2$

$$k_{\varkappa_2}(t) = t^{\varkappa_1 - \varkappa_2} k_{\varkappa_2}(t), \quad t > 0,$$

$t^{-\gamma}, t > 0, \gamma > 0$, are completely monotone functions and the complete monotonicity is preserved under multiplication, from Definition 3.11 it follows that $T_{\varkappa_1}(R_+) \subset T_{\varkappa_2}(R_+)$.

Closedness of $T_{\varkappa}(R_+)$ under convolutions and weak limits follows from the well-known properties that the complete monotonicity is preserved under formation of linear combinations and pointwise limits (see, e.g., [12]).

(ii) Observe that the characteristics and the Laplace exponent of $Tw_{\frac{\varkappa+1}{\varkappa}}(\alpha, \lambda)$ are equal

$$\left(0, \frac{\lambda^{\varkappa}\varkappa^{1+\varkappa}}{\Gamma(1+\varkappa)} t^{-2+\varkappa} e^{-\varkappa\lambda\alpha^{-\frac{1}{\varkappa}} t} \mathrm{d}t\right)$$

and

$$\psi_{\frac{1+\varkappa}{\varkappa},\alpha,\lambda}(\theta) = \frac{\lambda\varkappa}{\varkappa^{1-\varkappa}(\varkappa-1)}\left[\alpha^{\frac{\varkappa-1}{\varkappa}}\varkappa^{1-\varkappa} - \left(\varkappa\alpha^{-\frac{1}{\varkappa}} + \frac{\theta}{\lambda}\right)^{1-\varkappa}\right] \tag{3.12}$$

Because for each $\theta \geq 0$

$$\begin{aligned}\lim_{\varkappa\to\infty} \psi_{\frac{1+\varkappa}{\varkappa},\alpha,\lambda}(\theta) &= \lim_{\varkappa\to\infty} \frac{\lambda\varkappa\alpha^{\frac{\varkappa-1}{\varkappa}}}{\varkappa-1}\left[1 - \left(\frac{\theta}{\varkappa\lambda}\alpha^{\frac{1}{\varkappa}}\right)\right]^{1-\varkappa} \\ &= \alpha\lambda\left(1 - e^{-\frac{\theta}{\lambda}}\right),\end{aligned} \tag{3.13}$$

it follows that for all $\alpha > 0, \lambda > 0$ the scaled Poisson distributions $Tw_1(\alpha, \lambda) \in T_{\infty}(R_+)$.

Hawing in mind properties (i), we conclude from (3.13) that $T_{\infty}(R_+) = ID(R_+)$.

(iii) The case $\varkappa = \infty$ is contained in (ii).

Let $0 < \varkappa < \infty$. The statement (iii) follows easily from the Proposition 3.13, the formula (3.12) and the statement (i). □

Remark 3.16 Because

$$Tw_2(\alpha, \lambda)(\mathrm{d}t) = \frac{(\lambda\alpha^{-1})^{\lambda}}{\Gamma(\lambda)} t^{\lambda-1} e^{-\lambda\alpha^{-1}t} 1_{(0,\infty)}\mathrm{d}t, \quad \text{therefore} \quad T_1(R_+) = GGC.$$

Because

$$Tw_{\frac{3}{2}}(\alpha, \lambda)(C) = e^{-2\lambda\sqrt{\alpha}}\varepsilon_0(C) + \int_C \frac{2\lambda}{\sqrt{u}} e^{-2\frac{\lambda}{\sqrt{\alpha}}(u+\alpha)} I_1(4\alpha\sqrt{u})\mathrm{d}u, \quad C \in \mathscr{B}(R_+),$$

where $I_{\gamma}(z)$ is the modified Bessel function of the first kind (see Appendix), i.e.

$$I_{\gamma}(z) = \sum_{k=0}^{\infty} \frac{\left(\frac{z}{2}\right)^{2k+\gamma}}{k!\Gamma(\gamma+k+1)}, \quad \gamma \geq -1,$$

is the compound Poisson-exponential distribution, therefore $T_2(R_+)$ is the class of generalized convolutions of compound Poisson-exponential distributions, which coincides with the generalized mixed exponential convolutions, studied by Goldie [17], Steutel [18, 19] and Bondesson [11].

Example 3.17 (noncentral gamma distribution) Following Fisher [20] (see also [21, 22]) we say that $\Gamma_{\beta,\gamma,\lambda}$ is a noncentral gamma distribution with the shape parameter $\beta > 0$, the scale parameter $\gamma > 0$ and the noncentrality parameter $\lambda > 0$ if its pdf $f_{\beta,\gamma,\lambda}$ is the Poisson mixture of the gamma densities:

$$f_{\beta,\gamma,\lambda}(x) = e^{-\lambda} \sum_{j=0}^{\infty} \frac{\lambda^j}{j!} \frac{\beta^{\gamma+j} x^{\gamma+j-1}}{\Gamma(\gamma+j)} e^{-\beta x}$$
$$= e^{-\lambda-\beta x} \beta \left(\frac{\beta x}{\lambda}\right)^{\frac{\gamma-1}{2}} I_{\gamma-1}\left(\sqrt{\beta\lambda x}\right), \quad x > 0.$$

Fisher in [20] derived that the probability law

$$\mathscr{L}\left(\sum_{j=1}^{n} X_j^2\right) = \Gamma_{\frac{1}{2},\frac{n}{2},\lambda},$$

where $X_1, \ldots, X_n$ are independent, $\mathscr{L}(X_j) = N(\alpha_j, 1)$ and $\lambda = \frac{1}{2}\sum_{j=1}^{n} \alpha_j^2$.

Let $\mathrm{Bess}_{\beta,\lambda}$, $\beta > 0$, $\lambda > 0$, be a probability distribution on R_+, defined by the formula:

$$\mathrm{Bess}_{\beta,\lambda}(\mathrm{d}x) = e^{-\lambda}\varepsilon_0(\mathrm{d}x) + \beta e^{-\lambda-\beta x} I_1\left(2\sqrt{\beta\lambda x}\right) \mathrm{d}x.$$

Because

$$\int_0^{\infty} e^{-\theta x} f_{\beta,\gamma,\lambda}(x)\mathrm{d}x = e^{-\lambda}\beta(\beta\lambda)^{\gamma-1} \sum_{j=0}^{\infty} e^{-(\beta+\theta)x} \frac{(\beta\lambda x)^j}{j!\Gamma(\gamma+j)}$$
$$= e^{-\lambda}\left(\frac{\beta}{\theta+\beta}\right)^{\gamma} \sum_{j=0}^{\infty} \frac{1}{j!}\left(\frac{\beta\lambda}{\theta+\beta}\right)^j$$
$$= \left(\frac{\beta}{\theta+\beta}\right)^{\gamma} e^{\frac{-\lambda\theta}{\theta+\beta}} \tag{3.14}$$

and

$$\int_0^\infty e^{-\theta x}\mathrm{Bess}_{\beta,\lambda}(\mathrm{d}x) = e^{-\lambda} + \lambda\beta e^{-\lambda}\sum_{k=0}^{\infty}\int_0^\infty e^{-(\theta+\beta)x}\frac{(\beta\lambda x)^k}{k!(k+1)!}\mathrm{d}x$$
$$= e^{-\lambda} + e^{-\lambda}\sum_{k=0}^{\infty}\left(\frac{\beta\lambda}{\theta+\beta}\right)^{k+1}\frac{1}{(k+1)!} = e^{\frac{-\lambda\theta}{\theta+\beta}},$$

we find that

$$\Gamma_{\beta,\gamma,\lambda} = \Gamma_{\beta,\gamma} * \mathrm{Bess}_{\beta,\lambda},$$

implying equalities:

$$\Gamma_{\beta,\gamma_1,\lambda_1} * \Gamma_{\beta,\gamma_2,\lambda_2} = \Gamma_{\beta,\gamma_1+\gamma_2,\lambda_1+\lambda_2},$$
$$\Gamma_{\beta,\gamma_1,\lambda} * \Gamma_{\beta,\gamma_2} = \Gamma_{\beta,\gamma_1+\gamma_2,\lambda}$$

and

$$\Gamma_{\beta,\gamma,\lambda_1} * \mathrm{Bess}_{\beta,\lambda_2} = \Gamma_{\beta,\gamma,\lambda_1+\lambda_2}.$$

From (3.14) it follows that

$$\int_0^\infty e^{-\theta x} f_{\beta,\gamma,\lambda}(x)\mathrm{d}x = \exp\left\{\int_0^\infty \left(e^{-\theta u}-1\right)\left(\frac{\gamma}{u}+\lambda\right)e^{-\beta u}\mathrm{d}u\right\},$$

proving that the noncentral gamma distributions are infinitely divisible on R_+ with characteristics $(0, l_{\beta,\gamma,\lambda}(u)\mathrm{d}u)$, where

$$l_{\beta,\gamma,\lambda}(u) = \left(\frac{\gamma}{u}+\lambda\right)e^{-\beta u}, \quad u>0.$$

This function is completely monotone, implying that $\Gamma_{\beta,\gamma,\lambda} \in T_2(R_+)$. Because the function

$$k_{\beta,\gamma,\lambda}(u) := u l_{\beta,\gamma,\lambda}(u) = (\gamma+\lambda u)e^{-\beta u}, \quad u>0 \tag{3.15}$$

is not completely monotone, $\Gamma_{\beta,\gamma,\lambda}\bar{\in}T_1(R_+)$. From (3.15) it follows that $k_{\beta,\gamma,\lambda}$ is nondecreasing if and only if $\lambda \le \beta\gamma$. Only in this case the noncentral gamma distribution $\Gamma_{\beta,\gamma,\lambda}$ is self-decomposable.

Inverse noncentral gamma distribution

$$I\Gamma_{\beta,\gamma,\lambda}(\mathrm{d}x) := x^{-2} f_{\beta,\gamma,\lambda}(x^{-1})\mathrm{d}x$$

permits to define noncentral Student's t-distribution $T_d(\nu, \Sigma, \alpha, \lambda)$ with $\nu > 0$ degrees of freedom, a scaling matrix Σ, a location vector $\alpha \in R^d$, and a noncentrality parameter $\lambda > 0$ by means of the pdf $f_{\nu,\Sigma,\lambda}(x - \alpha)$, $x \in R^d$, where

$$
\begin{aligned}
f_{\nu,\Sigma,\lambda}(x) &= \int_0^\infty g_{0,u\Sigma}(x) u^{-2} f_{\frac{\nu}{2},\frac{\nu}{2},\lambda}\left(\frac{1}{u}\right) \mathrm{d}u \\
&= \int_0^\infty g_{0,u\Sigma}(x) e^{-\lambda} \sum_{j=0}^\infty \frac{\lambda^j}{j!} \frac{\left(\frac{\nu}{2}\right)^{\frac{\nu}{2}+j}}{\Gamma\left(\frac{\nu}{2}+j\right)} u^{-\frac{\nu}{2}-j-1} e^{-\frac{\nu}{2u}} \mathrm{d}u \\
&= \frac{e^{-\lambda}\left(\frac{\nu}{2}\right)^{\frac{\nu}{2}}}{(2\pi)^{\frac{d}{2}}\sqrt{|\Sigma|}} \left(\frac{\nu}{2} + \frac{1}{2}\langle x\Sigma^{-1}, x\rangle\right)^{-\frac{\nu+d}{2}} \\
&\quad \times \sum_{j=0}^\infty \frac{\Gamma\left(\frac{\nu+d}{2}+j\right)}{j!\Gamma\left(\frac{\nu}{2}+j\right)} \left(\frac{\lambda\nu}{\nu + \langle x\Sigma^{-1}, x\rangle}\right)^j \\
&= \frac{e^{-\lambda}}{(\nu\pi)^{\frac{d}{2}}\sqrt{|\Sigma|}} \left(\frac{\nu + \langle x\Sigma^{-1}, x\rangle}{\nu}\right)^{-\frac{\nu+d}{2}} \\
&\quad \times \sum_{j=0}^\infty \frac{\Gamma\left(\frac{\nu+d}{2}+j\right)}{j!\Gamma\left(\frac{\nu}{2}+j\right)} \left(\frac{\lambda\nu}{\nu + \langle x\Sigma^{-1}, x\rangle}\right)^j .
\end{aligned}
$$

Analogously we define doubly noncentral Student's t-distributions $T_d(\nu, \Sigma, \alpha, a, \lambda)$ with $\nu > 0$ degrees of freedom, a scaling matrix Σ, a location vector $\alpha \in R^d$, a noncentrality vector $a \in R^d$, and parameter $\lambda > 0$ by means of pdf $f_{\nu,\Sigma,a,\lambda}(x - \alpha)$, $x \in R^d$, where

$$
\begin{aligned}
f_{\nu,\Sigma,a,\lambda}(x) &= \int_0^\infty g_{ua,u\Sigma}(x) u^{-2} f_{\frac{\nu}{2},\frac{\nu}{2},\lambda}\left(\frac{1}{u}\right) \mathrm{d}u \\
&= \int_0^\infty g_{ua,u\Sigma}(x) e^{-\lambda} \sum_{j=0}^\infty \frac{\lambda^j}{j!} \frac{\left(\frac{\nu}{2}\right)^{\frac{\nu}{2}+j}}{\Gamma\left(\frac{\nu}{2}+j\right)} u^{-\frac{\nu}{2}-j-1} e^{-\frac{\nu}{2u}} \mathrm{d}u \\
&= \frac{e^{-\lambda}\left(\frac{\nu}{2}\right)^{\frac{\nu}{2}} 2\exp\left\{\langle x\Sigma^{-1}, x\rangle\right\}}{(2\pi)^{\frac{d}{2}}\sqrt{|\Sigma|}} \left(\frac{\langle a\Sigma^{-1}, a\rangle}{\nu + \langle x\Sigma^{-1}, x\rangle}\right)^{\frac{\nu+d}{4}} \\
&\quad \times \sum_{j=0}^\infty \frac{\left(\frac{\lambda\nu}{2}\right)^j}{j!\Gamma\left(\frac{\nu}{2}+j\right)} \left(\frac{\langle a\Sigma^{-1}, a\rangle}{\nu + \langle x\Sigma^{-1}, x\rangle}\right)^{\frac{j}{2}} \\
&\quad \times K_{\frac{\nu+d}{2}+j}\left(\left[\langle a\Sigma^{-1}, a\rangle\left(\nu + \langle a\Sigma^{-1}, x\rangle\right)\right]^{\frac{1}{2}}\right).
\end{aligned}
$$

Most likely, distributions $I\Gamma_{\beta,\gamma,\lambda}$, $T_d(\nu, \Sigma, \alpha, \lambda)$ and $T_d(\nu, \Sigma, \alpha, a, \lambda)$ are not infinitely divisible and do not correspond to any Lévy processes.

Example 3.18 (generalized gamma distribution). Recall that Bondesson introduced and studied in [11] a remarkable subclass of GGC of pdf on $(0, \infty)$, called the hyperbolically completely monotone pdf (HCM for short). It is said that f is HCM, if for every $u > 0$, $f(uv)f\left(\frac{u}{v}\right)$ is the completely monotone function in $w = v + v^{-1}$. For instance, GIG densities are HCM, because

$$gig(uv; \lambda, \chi, \psi)gig\left(\frac{u}{v}; \lambda, \chi, \psi\right) = \frac{\left(\frac{\psi}{\chi}\right)^{\lambda} u^{2(\lambda-1)}}{\left(2K_{\lambda}(\sqrt{\chi\psi})\right)^2} \exp\left\{-\frac{1}{2}\left(\chi u^{-1} + \psi u\right) w\right\}$$

and the function e^{-ax}, $x > 0$, $a > 0$, is, obviously, completely monotone.

The generalized gamma density functions $g_{\beta,\gamma,\delta}$ are defined by the formula (see, e.g., [11]):

$$g_{\beta,\gamma,\delta}(x) = \frac{|\delta|}{\Gamma(\gamma)}\beta^{\gamma} x^{\delta\gamma-1} \exp\left\{-\beta x^{\delta}\right\}, \quad x > 0, \quad \delta \in R_0^1, \quad \beta > 0, \quad \gamma > 0.$$

It is proved in [11] that for $0 < |\delta| \le 1$ $g_{\beta,\gamma,\delta}$ are HCM, because

$$g_{\beta,\gamma,\delta}(uv)g_{\beta,\gamma,\delta}\left(\frac{u}{v}\right) = \left(\frac{|\delta|\beta^{\gamma}}{\Gamma(\gamma)}\right)^2 u^{2(\delta\gamma-1)} \exp\left\{-\beta u^{\delta}(v^{\delta} + v^{-\delta})\right\}$$

and

$$\frac{\mathrm{d}}{\mathrm{d}w}\left(v^{\delta} + v^{-\delta}\right) = \frac{\delta \sin(\delta\pi)}{\pi} \int_{-\infty}^{0} \frac{|t|^{\delta}}{1 + t^2 - tw}\mathrm{d}t.$$

The statement now follows from the known properties of completely monotone functions (see, e.g., [12]).

For $\delta > 1$, pdf $g_{\beta,\gamma,\delta}$ are not infinitely divisible (see [11]) and, for $\delta < -1$, it is unknown whether or not $g_{\beta,\gamma,\delta}$ are infinitely divisible.

Following Definitions 1.1, 1.2 and using densities $g_{\beta,\gamma,\delta}$, $\delta < 0$, $\beta > 0$, $\gamma > 0$, it is natural to define the generalized Student's t-distributions with pdf as mixtures

$$\int_{0}^{\infty} g_{0,u\Sigma}(x - \alpha)g_{\beta,\gamma,\delta}(u)\mathrm{d}u, \quad x \in R^d$$

and the generalized noncentral Student's t-distributions with pdf as mixtures

$$\int_0^\infty g_{ua,u\Sigma}(x-\alpha)g_{\beta,\gamma,\delta}(u)\mathrm{d}u, \quad x \in R^d.$$

In the case $-1 \le \delta < 0$ their pdf are infinitely divisible, but, excepting $\delta = -1$, their Lévy measure had no tractable expressions.

3.4 Subordinated Lévy Processes

Subordination of Markov processes as a transformation through random time change was introduced by Bochner in 1949 (see [23, 24]). In the context of Lévy processes subordination give us possibility to construct and investigate statistical models with desirable feature of the marginal distributions.

Let $X = \{X_t, t \ge 0\}$ be a Lévy process in R^d, $X_0 \equiv 0$, with the triplet of Lévy characteristics (a, A, Π) and the characteristic exponent

$$\begin{aligned}\varphi(z) &:= -\log Ee^{i\langle z, X_t\rangle}\\ &= -i\langle a, z\rangle + \frac{1}{2}\langle zA, z\rangle - \int_{R_0^d}\left(e^{i\langle z,x\rangle} - 1 - i\langle z, x\rangle 1_{\{|x|\le 1\}}\right)\Pi(\mathrm{d}x), \quad z \in R^d,\end{aligned}$$

called the subordinand process.

Let $T = \{T_t, t \ge 0\}$ be a Lévy subordinator, $T_0 \equiv 0$, with the Laplace exponent

$$\psi(\theta) := -\log \mathrm{E}e^{-\theta T_1} = \beta_0\theta + \int_0^\infty \left(1 - e^{-\theta x}\right)\rho(\mathrm{d}x), \quad \theta \ge 0,$$

and characteristics (β_0, ρ), independent of X.

The subordinated process $\tilde{X} = \{\tilde{X}_t, t \ge 0\}$ is defined as a superposition

$$\tilde{X}_t = X_{T_t}, \quad t \ge 0.$$

The following theorem is obtained by Zolotarev [25], Bochner [24], Ikeda and Watanabe [28], and Rogozin [26]. It was treated by Feller [27] and Sato [2]. These ideas were extended to the multivariate subordination of Lévy processes by Barndorff-Nielsen et al. in 2001 (see [5]).

Let

$$\mu^t(B) = \mathrm{P}\{X_t \in B\}, \quad B \in \mathscr{B}(R^d), \quad t \geq 0,$$
$$\tau^t(C) = \mathrm{P}\{T_t \in C\}, \quad C \in \mathscr{B}(R_+), \quad t \geq 0$$

and

$$\tilde{\mu}^t(B) = \mathrm{P}\{\tilde{X}_t \in B\}, \quad B \in \mathscr{B}(R^d), \quad t \geq 0.$$

Theorem 3.19 (i) The subordinated process $\tilde{X} = \{\tilde{X}_t, t \geq 0\}$ is a Lévy process with characteristic exponent $\tilde{\varphi}(z) = \psi(\varphi(z)),\ z \in R^d$, and triplet of Lévy characteristics $(\tilde{a}, \tilde{A}, \tilde{\Pi})$, where

$$\tilde{a} = \beta_0 a + \int_{(0,\infty)} \left(\int_{|x| \leq 1} x \mu^s(dx) \right) \rho(ds),$$
$$\tilde{A} = \beta_0 A \tag{3.16}$$

and

$$\tilde{\Pi}(B) = \beta_0 \Pi(B) + \int_{(0,\infty)} \mu^s(B) \rho(ds), \quad B \in \mathscr{B}(R_0^d).$$

(ii) For $t \geq 0,\ B \in \mathscr{B}(R^d)$

$$\tilde{\mu}^t(B) = \int_{R_+} \mu^s(B) \tau^t(ds). \tag{3.17}$$

We refer the reader for proof to [2].

References

1. Appelbaum, D.: Lévy Processes and Stochastic Calculus. Cambridge University Press, Cambridge (2004)
2. Sato, K.: Lévy Processes and Infinitely Divisible Distributions. Cambridge University Press, Cambridge (1999)
3. Skorohod, A.V.: Random Processes with Independent Increments. Kluwer, Amsterdam (1991)
4. Rocha-Arteaga, A., Sato, K.: Topics in Infinitely Divisible Distributions and Lévy Processes. Sociedad Matematática Mexicana, Mexico (2003)
5. Barndorff-Niesen, O.E., Pedersen, J., Sato, K.: Multivariate subordination, self-decomposability and stability. Adv. Prob. **33**, 160–187 (2001)
6. Grigelionis, B.: Thorin classes of Lévy processes and their transforms. Lith. Math. J. **48**(3), 294–315 (2008)

7. Grigelionis, B.: On subordinated multivariate Gaussian Lévy processes. Acta Appl. Math. **96**, 233–249 (2007)
8. Sato, K., Yamazato, M.: Stationary processes of Ornstein-Uhlenbeck type. In: Probability Theory and Mathematical Statististics (Tbilisi, 1982), Lecture Notes in Mathematics, vol. 1021, pp. 541–551. Springer. Berlin (1983)
9. Bertoin, J.: Lévy Processes, Caml. Tracts. Math 1271, Cambridge University Press, Cambridge, (1996)
10. Bertoin, J.: Subordinators: Examples and applications. In: Ecole d'Ete de Probabilités de Saint-Flour XXVII, Lecture Notes in Math., vol. 1727, pp. 1–91. Springer. Heidelberg, (1999)
11. Bondesson, L.: Generalized gamma convolutions and related classes of distributions and densities. In: Lecture Notes in Statistics, vol. 76. Springer-Verlag, Berlin (1992)
12. Schilling, R., Song, R., Vondraček, Z.: Bernstein functions-theory and applications. De Gruyter (2010)
13. Jørgensen, B.: The Theory of Dispersion Models, Monographs on Statistics and Applied Probability, Vol. 76, Chapman & Hall, London (1997)
14. Tweedie, M.C.K.: An index which distinguishes between some important exponential families. In: Statistics: Applications and New Directions (Calcutta, 1981), Indian Statist. Inst., Calcutta, pp. 579–604 (1984)
15. Vinogradov, V.: Properties of certain Lévy and geometric Lévy processes. Commun. Stoch. Anal. **2**(2), 193–208 (2008)
16. Grigelionis, B.: Extending the Thorin class. Lith. Math. J. **51**(2), 194–206 (2011)
17. Goldie, C.: A class of infinitely divisible random variables. Proc. Cambridge Philos. Soc. **63**, 1141–1143 (1967)
18. Steutel, F.W.: Note on the infinite divisibility of exponential mixtures. Ann. Math. Statist. **38**, 1303–1305 (1967)
19. Steutel, F.W., Van Harn, K.: Infinite divisibility of probability distributions on the real line. Monographs and Textbooks in Pure and Appl. Math., vol. 259. Marcel Dekker, New York (2004)
20. Fisher, R.A.: The general sampling distribution of the multiple correlation coefficient. Proc. Royal Soc. London **121A**, 654–673 (1928)
21. Alam, K., Saxena, L.: Estimation of the noncentrality parameter of a chi-square distribution. Ann. Statist. **10**, 1012–1016 (1982)
22. Johnson, N.L., Kotz, S.: Distributions in Statistics: Continuous Multivariate Distributions. Wiley, New York (1972)
23. Bochner, S.: Diffusion Equation and Stochastic Processes. Proc. Nat. Acad. Sci. USA **35**, 368–370 (1949)
24. Bochner, S.: Harmonic analysis and theory of probability. Univ. California Press, Berkeley and Los Angeles (1955)
25. Zolotarev, V.M.: Distribution of the Superposition of Infinitely Divisible Processes. Probab. Theory Appl. **3**, pp. 185–188 (1958)
26. Rogozin. B.A.: On some class of processes with independent increments. Theory Probal. Appl. **10**, pp. 479–483 (1965)
27. Feller, W.: An Introduction to Probability Theory and its Applications. Vol. 2, 2nd edn. Wiley, New York (1971)
28. Ikeda, N., Watanabe, S.: On some relations between the harmonic measure and the Lévy measure for certain class of Markov processes. J. Math. Kyoto Univ. **2**, 79–95 (1962)

Chapter 4
Student-Lévy Processes

Important classes of Lévy processes as statistical models arise as the subordinated multivariate Gaussian Lévy process with a mean vector $a \in R^d$ and a non-degenerated covariance matrix A.

For instance, taking

$$\mu^t(B) = \int_B g_{ta,tA}(x)\mathrm{d}x, \quad t \geq 0, \quad B \in \mathscr{B}(R^d)$$

and

$$\tau^1(\mathrm{d}x) = gig(x; \lambda, \chi, \psi)\mathrm{d}x,$$

we shall obtain the famous class of generalized hyperbolic processes.

Having in mind Theorem 3.15, properties of Thorin subordinated multivariate Gaussian Lévy processes are of fundamental importance investigating many statistical models, including stochastic processes related to Student's t-distribution.

Theorem 4.1 *[1]. Let $X = \{X_t, t \geq 0\}$ be a Gaussian Lévy process in R^d with mean vector $a \in R^d$ and a non-degenerated covariance matrix A. Let $T^{(\varkappa)} = \{T_t^{(\varkappa)}, t \geq 0\}$ be an independent of X Thorin's subordinator with Laplace exponent $\psi_\varkappa(\theta)$, defined by the formulas* (3.10) *and* (3.11)*, and characteristics $(\beta_\varkappa, \rho_\varkappa)$, where*

$$\rho_\varkappa(\mathrm{d}t) = t^{\varkappa-2} \int_0^\infty e^{-vt} Q_\varkappa(\mathrm{d}v)\mathrm{d}t. \tag{4.1}$$

Then:

(i) the triplet of Lévy characteristics of the subordinated process

$$X^{(\varkappa)} = \{X_t^{(\varkappa)} := X_{T_t^{(\varkappa)}}, t \geq 0\}$$

B. Grigelionis, *Student's* t*-Distribution and Related Stochastic Processes*, SpringerBriefs in Statistics, DOI: 10.1007/978-3-642-31146-8_4,

equals $(a_\varkappa, A_\varkappa, \Pi_\varkappa)$, *where*

$$a_\varkappa = \beta_\varkappa a + \int\limits_{\{|x|\le 1\}} x l_\varkappa(x)\mathrm{d}x,$$

$$A_\varkappa = \beta_\varkappa A, \tag{4.2}$$

$$\Pi_\varkappa(B) = \int\limits_B l_\varkappa(x)\mathrm{d}x, \quad B \in \mathscr{B}(R_0^d),$$

$$l_\varkappa(x) = \frac{2\exp\{\langle aA^{-1}, x\rangle\}}{\sqrt{|A|}(2\pi)^{\frac{d}{2}}(\langle xA^{-1}, x\rangle)^{\frac{d}{2}+1-\varkappa}} \int\limits_0^\infty (h(v,x))^{\frac{d}{2}+1-\varkappa} K_{\frac{d}{2}+1-\varkappa}(h(v,x)) Q_\varkappa(\mathrm{d}v)$$

and

$$h(v,x) = [(\langle aA^{-1}, a\rangle + 2v)\langle xA^{-1}, x\rangle]^{\frac{1}{2}};$$

(ii) $X^{(\varkappa)}$ *is self-decomposable if and only if, for a.e.* $\xi \in S^{d-1}$ *with respect to the surface Lebesgue measure on* S^{d-1}*, the function* $k_\xi^{(\varkappa)}(r) := r^{\mathrm{d}} l_\varkappa(r\xi)$, $r > 0$, *is decreasing;*

(iii) if $d = 1$ *or* $d \ge 2$ *and* $a = 0$*, then* $X^{(1)}$ *is self-decomposable;*

(iv) if $d = 2$, $a \neq 0$ *and*

$$\int\limits_0^\infty (1+v)^2 Q_1(\mathrm{d}v) < \infty \tag{4.3}$$

or $d \ge 3$, $a \neq 0$, *and*

$$\int\limits_0^\infty (1+v) Q_1(\mathrm{d}v) < \infty, \tag{4.4}$$

then $X^{(1)}$ *is not self-decomposable;*

(v) if $\varkappa > 1$, $\varkappa \neq \frac{d}{2}$, *and*

$$\int\limits_0^\infty (1+v) Q_\varkappa(\mathrm{d}v) < \infty, \tag{4.5}$$

or $\varkappa > 1$, $\varkappa = \frac{d}{2}$, *and*

$$\int\limits_0^\infty (1+v)^2 Q_\varkappa(\mathrm{d}v) < \infty, \tag{4.6}$$

then $X^\varkappa$ *is not self-decomposable.*

Proof

(i) We shall use the following formulas:

$$\int_0^\infty t^{-\alpha-1}e^{-\gamma t-\frac{\delta}{t}}\mathrm{d}t = 2\left(\frac{\gamma}{\delta}\right)^{\frac{\alpha}{2}} K_\alpha\left(2\sqrt{\gamma\delta}\right), \tag{4.7}$$

$$\alpha > 0, \quad \gamma > 0, \quad \delta > 0 \quad (\text{see Appendix}),$$

and

$$\begin{aligned}
&\int_0^\infty g_{ta,tA}(x)t^{-\alpha-1}e^{-\gamma t-\frac{\delta}{t}}\mathrm{d}t \\
&= \frac{\exp\{\langle aA^{-1},x\rangle\}}{\sqrt{|A|}(2\pi)^{\frac{d}{2}}}\int_0^\infty t^{-\alpha-\frac{d}{2}-1} \\
&\times \exp\left\{-\left(\gamma+\frac{1}{2}\langle aA^{-1},a\rangle\right)t-\left(\delta+\frac{1}{2}\langle xA^{-1},x\rangle\right)t^{-1}\right\}\mathrm{d}t \\
&= \frac{2\exp\{\langle aA^{-1},x\rangle\}}{\sqrt{|A|}(2\pi)^{\frac{d}{2}}}\left(\frac{2\gamma+\langle aA^{-1},a\rangle}{2\delta+\langle xA^{-1},x\rangle}\right)^{\frac{\alpha}{2}+\frac{d}{4}} \\
&\times K_{\alpha+\frac{d}{2}}([(2\gamma+\langle aA^{-1},a\rangle)(2\delta+\langle xA^{-1},x\rangle)]^{\frac{1}{2}}), \\
&\gamma \geq 0, \quad \delta \geq 0, \quad |a| > 0, |x| > 0.
\end{aligned} \tag{4.8}$$

From formulas (3.16), (4.1), (4.8) and Theorem 3.19 we find that the statement (i) holds with the function

$$\begin{aligned}
l_\varkappa(x) &= \int_0^\infty g_{ta,tA}(x)t^{\varkappa-2}\int_0^\infty e^{-vt}Q_\varkappa(\mathrm{d}v)\mathrm{d}t \\
&= \frac{\exp\{\langle aA^{-1},x\rangle\}}{\sqrt{|A|}(2\pi)^{\frac{d}{2}}}\int_0^\infty\int_0^\infty t^{-\frac{d}{2}-2+\varkappa} \\
&\quad\times \exp\left\{-\frac{1}{2t}\langle xA^{-1},x\rangle-\frac{t}{2}[\langle aA^{-1},a\rangle+2v]\right\}\mathrm{d}t\,Q_\varkappa(\mathrm{d}v) \\
&= \frac{2\exp\{\langle aA^{-1},x\rangle\}}{\sqrt{|A|}(2\pi)^{\frac{d}{2}}}\int_0^\infty\int_0^\infty u^{-\frac{d}{2}-2+\varkappa}\left(\frac{2}{\langle aA^{-1},a\rangle+2v}\right)^{-\frac{d}{2}-1+\varkappa} \\
&\quad\times \exp\left\{-\frac{h(v,x)}{4u}-u\right\}\mathrm{d}u\,Q_\varkappa(\mathrm{d}v)
\end{aligned}$$

$$= \frac{2\exp\{\langle aA^{-1}, x\rangle\}}{\sqrt{|A|}(2\pi)^{\frac{d}{2}}\langle xA^{-1}, x\rangle^{\frac{d}{2}+1-\varkappa}}$$
$$\times \int_0^\infty (h(v,x))^{\frac{d}{2}+1-\varkappa} K_{\frac{d}{2}+1-\varkappa}(h(v,x)) Q_\varkappa(\mathrm{d}v).$$

(ii) This statement follows directly from Proposition 3.8, Theorem 3.9 and (i).
(iii) Since

$$k_\xi^{(1)}(r) = r^{\frac{d}{2}} \frac{2\exp\{r\langle aA^{-1}, \xi\rangle\}}{\sqrt{|A|}(2\pi)^{\frac{d}{2}}(\langle \xi A^{-1}, \xi\rangle)^{\frac{d}{2}}} \int_0^\infty (h(v,\xi))^{\frac{d}{2}} K_{\frac{d}{2}}(rh(v,\xi)) Q_1(\mathrm{d}v)$$

and (see Appendix)

$$K'_\gamma(z) = -\left(K_{\gamma-1}(z) + \frac{\gamma}{z}K_\gamma(z)\right),$$

we have that

$$\frac{\mathrm{d}}{\mathrm{d}r}k_\xi^{(1)}(r) = \frac{2\exp\{r\langle aA^{-1}, \xi\rangle\}}{\sqrt{|A|}(2\pi)^{\frac{d}{2}}(\langle \xi A^{-1}, \xi\rangle)^{\frac{d}{2}}} r^{\frac{d}{2}} \int_0^\infty (h(v,\xi))^{\frac{d}{2}}$$
$$\times [K_{\frac{d}{2}}(rh(v,\xi))\langle aA^{-1}, \xi\rangle - K_{\frac{d}{2}-1}(rh(v,\xi))\, h(v,\xi)] Q_1(\mathrm{d}v).$$

If $a = 0$, then, for all $\xi \in S^{d-1}$,

$$\frac{\mathrm{d}}{\mathrm{d}r}k_\xi^{(1)}(r) = -\frac{2r^{\frac{d}{2}}}{\sqrt{|A|}(2\pi)^{\frac{d}{2}}\left(\langle \xi A^{-1}, \xi\rangle\right)^{\frac{d}{2}}}$$
$$\times \int_0^\infty (h(v,\xi))^{\frac{d}{2}+1} K_{\frac{d}{2}-1}(rh(v,\xi))\, Q_1(\mathrm{d}v) < 0,$$

proving that, for al $\xi \in S^{d-1}$, the function $k_\xi^{(1)}(r)$, $r > 0$, is decreasing.
If $d = 1$, then, since

$$K_{\frac{1}{2}}(z) = K_{-\frac{1}{2}}(z),$$

for $\xi = \pm 1$, we have that

$$\frac{\mathrm{d}}{\mathrm{d}r} k_\xi^{(1)}(r) = \frac{-2\exp\{r\langle aA^{-1}, \xi\rangle\}}{\sqrt{2\pi}\langle \xi A^{-1}, \xi\rangle} \int_0^\infty \sqrt{rh(v,\xi)} K_{\frac{1}{2}}(rh(v,\xi)) \times \left\{ \left[\left(\langle aA^{-1}, a\rangle + 2v \right) \langle \xi A^{-1}, \xi\rangle \right]^{\frac{1}{2}} - \langle aA^{-1}, \xi\rangle \right\} Q_1(\mathrm{d}v) < 0$$

proving again that the function $k_\xi^{(1)}(r)$, $r > 0$, is decreasing.

(iv) If $d \geq 3$ and $a \neq 0$, since, for all $\gamma \neq 0$, as $v \downarrow 0$

$$v^\gamma K_\gamma(v) = \int_v^\infty w^\gamma K_{\gamma-1}(w)\mathrm{d}w \uparrow \Gamma(|\gamma|)2^{|\gamma|-1}, \tag{4.9}$$

we find that

$$\frac{\mathrm{d}}{\mathrm{d}r} k_\xi^{(1)}(r) = \frac{2\exp\left\{r\langle aA^{-1}, \xi\rangle\right\}}{\sqrt{|A|}(2\pi)^{\frac{d}{2}}\langle \xi A^{-1}, \xi\rangle} \times \left[\langle aA^{-1}, \xi\rangle \int_0^\infty \int_{rh(v,\xi)} w^{\frac{d}{2}} K_{\frac{d}{2}-1}(w)\mathrm{d}w Q_1(\mathrm{d}v) - r\int_0^\infty h^2(v,\xi) \int_{rh(v,\xi)} w^{\frac{d}{2}-1} K_{\frac{d}{2}-2}(w)\mathrm{d}w Q_1(\mathrm{d}v) \right]. \tag{4.10}$$

Under the assumption (4.4), for $\xi \in S^{d-1} \cap \left\{\xi : \langle aA^{-1}, \xi\rangle > 0\right\}$ and sufficiently small r, we have that

$$\frac{\mathrm{d}}{\mathrm{d}r} k_\xi^{(1)}(r) > 0,$$

showing that $k_\xi^{(1)}(r)$, $r > 0$, is nondecreasing in r for a subset on S^{d-1} of positive surface Lebesgue measure.

If $d = 2$ and $a \neq 0$, from Grosswald's formula we find that

$$K_0(v) = vK_1(v) \int_0^\infty \frac{g_1(u)}{v^2 + u} \mathrm{d}u,$$

where $vK_1(v) \uparrow 1$, as $v \downarrow 0$.

But (see [2])

$$g_1(t) = 2[\pi^2 t(J_1^2(\sqrt{t}) + Y_1^2(\sqrt{t}))]^{-1},$$
$$g_1(t) \sim t^{-\frac{1}{2}}, \quad \text{as} \quad t \to \infty, \quad \text{and} \quad g_1(t) \to 1, \quad \text{as} \quad t \to 0,$$

implying that

$$K_0(v) \leq \int_1^\infty \frac{g_1(u)}{u} \mathrm{d}u + \max_{0 \leq u \leq 1} g_1(u) \log\left(\frac{v^2+1}{v^2}\right). \tag{4.11}$$

Now from (4.9) and (4.10) derive that

$$\frac{\mathrm{d}}{\mathrm{d}r} k_\xi^{(1)}(r) = \frac{2\exp\left\{r\langle aA^{-1}, \xi\rangle\right\}}{\sqrt{|A|}2\pi\langle \xi A^{-1}, \xi\rangle} \times \left[\langle aA^{-1}, \xi\rangle \int_0^\infty \int_{rh(v,\xi)} wK_0(w)\mathrm{d}wQ_1(\mathrm{d}v) - \int_0^\infty rh^2(v,\xi)K_0(rh(v,\xi))\, Q_1(\mathrm{d}v)\right]. \tag{4.12}$$

Because $\log(v^2+1) \leq v^2$, from (4.6), (4.11), and (4.12) we again obtain that, for $\xi \in S^{d-1} \cap \left\{\xi : \langle aA^{-1}, \xi\rangle > 0\right\}$ and sufficiently small r,

$$\frac{\mathrm{d}}{\mathrm{d}r} k_\xi^{(1)}(r) > 0,$$

proving that $X^{(1)}$ is not self-decomposable.

(v) Since

$$k_\xi^{(\varkappa)}(r) := r^d g_\varkappa(r\xi) = \frac{2\exp\left\{r\langle aA^{-1}, \xi\rangle\right\} r^{2\varkappa-2}}{\sqrt{|A|}(2\pi)^{\frac{d}{2}}\left(\langle \xi A^{-1}, \xi\rangle\right)^{\frac{d}{2}+1-\varkappa}} \times \int_0^\infty \int_{rh(v,\xi)} w^{\frac{d}{2}+1-\varkappa} K_{\frac{d}{2}-\varkappa}(w)\mathrm{d}wQ_\varkappa(\mathrm{d}v),$$

$$\frac{\mathrm{d}}{\mathrm{d}r} k_\xi^{(\varkappa)}(r) = \frac{2\exp\left\{r\langle aA^{-1}, \xi\rangle r^{2\varkappa-2}\right\}}{\sqrt{|A|}(2\pi)^{\frac{d}{2}}\left(\langle \xi A^{-1}, \xi\rangle\right)^{\frac{d}{2}+1-\varkappa}}$$

$$\times \left[\int_0^\infty \int_{rh(v,\xi)} u^{\frac{d}{2}+1-\varkappa} K_{\frac{d}{2}-\varkappa}(u) Q_\varkappa(\mathrm{d}v) \left(\langle aA^{-1}, \xi \rangle + \frac{2\varkappa - 2}{r} \right) \right.$$
$$\left. - \int_0^\infty rh^2(v,\xi) \int_{rh(v,\xi)}^\infty u^{\frac{d}{2}-\varkappa} K_{\frac{d}{2}-\varkappa-1}(u) \mathrm{d}u \, Q_\varkappa(\mathrm{d}v) \right]. \tag{4.13}$$

Let $\varkappa > 1$, $\varkappa \neq \frac{d}{2}$. Using (4.9),

$$\int_{rh(v,\xi)}^\infty u^{\frac{d}{2}-\varkappa} K_{\frac{d}{2}-\varkappa-1}(u) \mathrm{d}u = (rh(v,\xi))^{\frac{d}{2}\varkappa}$$
$$\times K_{\frac{d}{2}-\varkappa}(rh(v,\xi)) \uparrow \Gamma\left(\left| \frac{d}{2} - \varkappa \right| \right) 2^{|\frac{d}{2}-\varkappa|-1}, \quad \text{as} \quad r \downarrow 0. \tag{4.14}$$

In this case from (4.13) and (4.14) under the assumption (4.5), for all $\xi \in S^{d-1}$ and sufficiently small r, we get that

$$\frac{\mathrm{d}}{\mathrm{d}r} k_\xi^{(\varkappa)}(r) > 0,$$

implying that $X^{(\varkappa)}$ is not self-decomposable. Finally, if $\varkappa > 1$, $\varkappa = \frac{d}{2}$, from (4.13) we have that

$$\frac{\mathrm{d}}{\mathrm{d}r} k_\xi^{\left(\frac{d}{2}\right)}(r) = \frac{2 \exp\left\{ r \langle aA^{-1}, \xi \rangle \right\} r^{d-2}}{\sqrt{|A|} (2\pi)^{\frac{d}{2}} \langle \xi A^{-1}, \xi \rangle}$$
$$\times \left[\int_0^\infty \int_{rh(v,\xi)}^\infty u K_0(u) \mathrm{d}u \, Q_{\frac{d}{2}}(\mathrm{d}v) \left(\langle aA^{-1}, \xi \rangle + \frac{d-2}{r} \right) \right.$$
$$\left. - \int_0^\infty rh^2(v,\xi) K_0(rh(v,\xi)) \, Q_{\frac{d}{2}}(\mathrm{d}v) \right]. \tag{4.15}$$

Now from (4.6), (4.11) and (4.15) we obtain that for all $\xi \in S^{d-1}$ and sufficiently small r

$$\frac{\mathrm{d}}{\mathrm{d}r} k_\xi^{\left(\frac{d}{2}\right)}(r) > 0,$$

proving that $X^{\left(\frac{d}{2}\right)}$ is not self-decomposable. □

Remark 4.2 The statement (iii) of Theorem 4.1 is contained in [3] and [4]. Some related results are obtained in [5, 6].

Definition 4.3 A d-dimensional Lévy process $X = \{X_t, t \geq 0\}$ is called the Student-Lévy process if

$$\mathscr{L}(X_1) = T_d(\nu, \Sigma, \alpha).$$

Definition 4.4 A d-dimensional Lévy process $X^{(a)} = \{X_t^{(a)}, t \geq 0\}$ is called the noncentral Student-Lévy process with the noncentrality vector $a \in R_0^d$ if

$$\mathscr{L}(X_1^{(a)}) = T_d(\nu, \Sigma, \alpha, a).$$

Proposition 4.5

(i) The Student-Lévy process $X = \{X_t, t \geq 0\}$ has the following structure:

$$X_t = G_{T_t} + \alpha t, \quad t \geq 0,$$

where $G = \{G_t, t \geq 0\}$ is a Gaussian Lévy process with the triplet $(0, \Sigma, 0)$ of Lévy characteristics and $T = \{T_t, t \geq 0\}$ is an independent of G Lévy subordinator such that

$$\mathscr{L}(T_1) = GIG\left(-\frac{\nu}{2}, \nu, 0\right). \tag{4.16}$$

(ii) The triplet of Lévy characteristics of X equals $(\gamma_0, 0, \Pi_0)$, where

$$\gamma_0 = \int_{\{|x| \leq 1\}} x l_0(x) \mathrm{d}x + \alpha,$$

$$\Pi_0(B) = \int_B l_0(x) \mathrm{d}x, \quad B \in \mathscr{B}(R_0^d),$$

and

$$l_0(x) = \frac{\nu 2^{\frac{d}{4}+1} (\langle x\Sigma^{-1}, x\rangle)^{-\frac{d}{4}}}{\sqrt{|\Sigma|}(2\pi)^{\frac{d}{2}}} \int_0^\infty u^{\frac{d}{4}} K_{\frac{d}{2}}\left(\left(2t\langle x\Sigma^{-1}, x\rangle\right)^{\frac{1}{2}}\right) g_{\frac{\nu}{2}}(2\nu t) \mathrm{d}t.$$

(iii) X is self-decomposable.

Proof (i) It is enough to observe that from (4.16) it follows the equality

$$\mathscr{L}(G_{T_1}+\alpha)=T_d(\nu,\Sigma,\alpha).$$

(ii) From formulas (1.4) and (4.16) we have that $GIG\left(-\frac{\nu}{2},\nu,0\right)\in T_1(R_+)$ with zero drift and the Thorin measure $\nu g_{\frac{\nu}{2}}(2\nu t)\mathrm{d}t$. Now from Theorem 4.1 we find that X has the triplet of Lévy characteristics $(\gamma_0,0,\Pi_0)$, where

$$\gamma_0=\int\limits_{\{|x|\le 1\}} x l_0(x)\mathrm{d}x,$$

$$\Pi_0(B)=\int\limits_B l_0(x)\mathrm{d}x,\quad B\in\mathscr{B}(R_0^d)$$

and

$$\begin{aligned}l_0(x)&=\frac{2\left(\langle x\Sigma^{-1},x\rangle\right)^{-\frac{d}{2}}}{\sqrt{|\Sigma|}(2\pi)^{\frac{d}{2}}}\\&\quad\times\int\limits_0^\infty\left(2t\langle x\Sigma^{-1},x\rangle\right)^{\frac{d}{4}}K_{\frac{d}{2}}\left(\left(2t\langle x\Sigma^{-1},x\rangle\right)^{\frac{1}{2}}\right)\nu g_{\frac{\nu}{2}}(2\nu t)\mathrm{d}t\\&=\frac{\nu 2^{\frac{\nu}{4}+1}\left(\langle x\Sigma^{-1},x\rangle\right)^{-\frac{d}{4}}}{\sqrt{|\Sigma|}(2\pi)^{\frac{d}{2}}}\int\limits_0^\infty t^{\frac{d}{4}}K_{\frac{d}{2}}\left(\left(2t\langle x\Sigma^{-1},x\rangle\right)^{\frac{1}{2}}\right)g_{\frac{\nu}{2}}(2\nu t)\mathrm{d}t.\end{aligned}$$

(iii) The statement following directly from Theorem 4.1 (iii). □

Proposition 4.6 *(i) The noncentral Student-Lévy process $X^{(a)}=\{X_t^{(a)},t\ge 0\}$ with the noncentrality vector $a\in R_0^d$ has the following structure:*

$$X_t^{(a)}=G_{T_t}^{(a)}+\alpha t,\quad t\ge 0,$$

where $G^{(a)}=\left\{G_t^{(a)},t\ge 0\right\}$ is a Gaussian Lévy process with the triplet $(a,\Sigma,0)$ of Lévy characteristics and $T=\{T_t,t\ge 0\}$ is an independent of $G^{(a)}$ Lévy subordinator such that

$$\mathscr{L}(T_1)=GIG\left(-\frac{\nu}{2},\nu,0\right).$$

(ii) The triplet of Lévy characteristics of $X^{(a)}$ equals $(\gamma_a,0,\Pi_a)$, where

$$\gamma_a = \int_{\{|x|\le 1\}} x l_a(x)\mathrm{d}x,$$

$$\Pi_a(B) = \int_B l_a(x)\mathrm{d}x, \quad B \in \mathscr{B}(R^d)$$

and

$$l_a(x) = \frac{2\nu \exp\left\{\langle a\Sigma^{-1}, x\rangle\right\}}{\sqrt{|\Sigma|}(2\pi)^{\frac{d}{2}} \left(\langle x\Sigma^{-1}, x\rangle\right)^{\frac{d}{4}}} \int_0^\infty \left(\langle a\Sigma^{-1}, a\rangle + 2t\right)^{\frac{d}{4}} \times K_{\frac{d}{2}}\left(\left(\langle a\Sigma^{-1}, a\rangle + 2t\right)\langle x\Sigma^{-1}, x\rangle\right)^{\frac{1}{2}} g_{\frac{\nu}{2}}(2\nu t)\mathrm{d}t.$$

(iii) *$X^{(a)}$ is self-decomposable if and only if the function $r^d l^{(a)}(r\xi)$, $r > 0$, is decreasing for a.e. $\xi \in S^{d-1}$ with respect to the surface Lebesgue measure on S^{d-1}.*

(iv) *If $d = 1$, $X^{(a)}$ is self-decomposable.*

Proof (i) It is enough to observe that from (4.16) it follows the equality

$$\mathscr{L}(G^{(a)}_{T_1} + \alpha) = T_d(\nu, \Sigma, \alpha, a).$$

(ii) From formulas (1.4) and (4.16) we have that $GIG\left(-\frac{\nu}{2}, \nu, 0\right) \in T_1(R_+)$ with zero drift and the Thorin's measure $\nu g_{\frac{\nu}{2}}(2\nu t)\mathrm{d}t$. Now the statement follows directly from the Theorem 4.1 (i).

(iii) The statement follows directly from Theorem 4.1 (ii).

(iv) The statement is the corollary of Theorem 4.1 (iii). □

References

1. Grigelionis, B.: Thorin classes of Lévy processes and their transforms. Lith. Math. J. **48**(3), 294–315 (2008)
2. Watson, G.N.: Treatise on the theory of Bessel functions. Cambridge Univ. Press, Cambridge (1958)
3. Barndorff-Niesen, O.E., Pedersen, J., Sato, K.: Multivariate subordination, self-decomposability and stability. Adv. Prob. **33**, 160–187 (2001)
4. Sato, K.: Subordination and self-decomposability. Stat. Probab. Lett. **54**, 317–324 (2001)
5. Takano, K.: On mixtures of the normal distribution by the generalized gamma convolutions. Bull. Fac. Sci. Ibaraki Univ. **21**, 29–41 (1989)
6. Takano, K.: On Bessel equations and the Lévy representation of the multivariate t-distribution. Technical report, Department of Mathematics, Ibaraki University, Japan (1994)

Chapter 5
Student OU-Type Processes

The classical Ornstein–Uhlenbeck process $\{X_t, t \geq 0\}$, starting from $x \in R^d$, is a solution of linear equation.

$$X_t = x + B_t - c\int_0^t X_s \mathrm{d}s, \quad t \geq 0, \tag{5.1}$$

where $c > 0$ and $\{B_t, t \geq 0\}$ is a standard d-dimensional Brownian motion. It is uniquely solved by

$$X_t = e^{-ct}x + \int_0^t e^{-c(t-s)}\mathrm{d}B_s, \quad t \geq 0,$$

where the last integral is a Wiener stochastic integral. We easily find that

$$\mathscr{L}(X_t) = G_{e^{-ct}x, \frac{1}{2c}(1-e^{-2ct})I_d} \Rightarrow G_{0, \frac{1}{2c}I_d},$$

as $t \to \infty$, where I_d is an identity $d \times d$ matrix.

If we replace $\{B_t, t \geq 0\}$ in (5.1) by an arbitrary Lévy process $\{Z_t, t \geq 0\}$ with the triplet (a, A, Π) of Lévy characteristics and the characteristic exponent $\varphi(z) = -\log Ee^{i\langle z, X_1\rangle}$, the solution

$$X_t = e^{-ct}x + \int_0^t e^{-c(t-s)}\mathrm{d}Z_s, \quad t \geq 0 \tag{5.2}$$

is called the starting from $x \in R^d$ Ornstein–Uhlenbeck type process generated by (a, A, Π, c).

B. Grigelionis, *Student's* t*-Distribution and Related Stochastic Processes*,
SpringerBriefs in Statistics, DOI: 10.1007/978-3-642-31146-8_5,

The integral in (5.2) is defined analogously to the Wiener integral through converging in probability integral sums (see, e.g., [1]).

If we write

$$P_t(x, B) = \mathrm{P}\{X_t \in B\}, \quad x \in R^d, \quad B \in \mathscr{B}(R^d), \quad t \geq 0,$$

it can be proved (see [2]) that

$$\int_{R^d} e^{i\langle z,y\rangle} P_t(x, \mathrm{d}y) = \exp\left\{ ie^{-ct}\langle x, z\rangle - \int_0^t \varphi(e^{-cs}z)\mathrm{d}s \right\}, \quad x, z \in R^d, \quad t \geq 0,$$

implying that $P_t(x, \cdot)$ is an infinitely divisible probability measure with the triplet $(a_{t,x}, A_t, \Pi_t)$ of Lévy characteristics given by the formulas:

$$A_t = \int_0^t e^{-2cs}\mathrm{d}s A, \quad t \geq 0$$

$$\Pi_t(B) = \int_{R_0^d}\int_0^t 1_B(e^{-cs}y)\mathrm{d}s\,\Pi(\mathrm{d}y), \quad B \in \mathscr{B}(R_0^d), \quad t \geq 0,$$

and

$$\begin{aligned} a_{t,x} &= e^{-ct}x + \int_0^t e^{-cs}\mathrm{d}s \\ &\quad + \int_{R_0^d} \int_0^t e^{-cs}y\left(1_{\{e^{-cs}|y|\leq 1\}} - 1_{\{|y|\leq 1\}}\right)\mathrm{d}s\,\Pi(\mathrm{d}y), \quad t \geq 0, \quad x \in R^d. \end{aligned}$$

Because

$$\begin{aligned} &\int_{R^d}\int_{R^d} e^{i\langle z,w\rangle} P_s(y, \mathrm{d}w)P_t(x, \mathrm{d}y) \\ &\quad = \int_{R^d} \exp\left\{ i\langle y, e^{-cs}z\rangle - \int_0^s \varphi(e^{-cr}z)\mathrm{d}r \right\} P_t(x, \mathrm{d}y) \\ &\quad = \exp\left\{ i\langle x, e^{-c(t+s)}z\rangle - \int_0^s \varphi(e^{-c(r+s)}z)\mathrm{d}r - \int_0^t \varphi(e^{-cr}z)\mathrm{d}r \right\} \end{aligned}$$

$$= \int_{R^d} e^{i\langle z,w\rangle} P_{t+s}(x, \mathrm{d}w),$$

$P_t(x, B), t \geq 0, x \in R^d, B \in \mathscr{B}(R^d)$, satisfies the Chapman–Kolmogorov identity

$$\int_{R^d} P_t(x, \mathrm{d}y) P_s(y, B) = P_{t+s}(x, B)$$

as the transition probability function of the time homogeneous Markov process X.

It is known (see [2–6]) that, as $t \to \infty$, for each $x \in R^d$

$$P_t(x, \cdot) \Rightarrow \tilde{\mu}_c$$

if and only if

$$\int_{\{|y|>2\}} \log |y| \Pi(\mathrm{d}y) < \infty, \tag{5.3}$$

where the limit distribution $\tilde{\mu}_c$ satisfies

$$\int_{R^d} e^{i\langle z,y\rangle} \tilde{\mu}_c(\mathrm{d}y) = \exp\left\{-\int_0^\infty \varphi(e^{-cs}z)\mathrm{d}s\right\}, \quad z \in R^d. \tag{5.4}$$

The distribution $\tilde{\mu}_c$ is self-decomposable with the triplet of Lévy characteristics $(\tilde{a}_c, \tilde{A}_c, \tilde{\Pi}_c)$, where

$$\tilde{a}_c = \frac{1}{c}a + \frac{1}{c} \int_{\{|y|>1\}} \frac{y}{|y|} \Pi(\mathrm{d}y),$$

$$\tilde{A}_c = \frac{1}{2c} A$$

and

$$\tilde{\Pi}_c(B) = \frac{1}{c} \int_{R^d} \int_0^\infty 1_B \left(e^{-s}y\right) \mathrm{d}s \Pi(\mathrm{d}y), \quad B \in \mathscr{B}(R_0^d).$$

There is one-to-one continuous in the topology of weak convergence correspondence between the class $ID_{\log}(R^d)$ of infinitely divisible distributions, satisfying the integrability assumption (5.3), and the class of self-decomposable distributions $L(R^d)$. It is given by the mapping

$$ID_{\log}(R^d) \ni \mu = \mathscr{L}(Z_1) \leftrightarrow \mathscr{L}\left(\int_0^\infty e^{-t}\mathrm{d}Z_t\right) = \tilde{\mu} \in L(R^d). \tag{5.5}$$

The correspondence (5.5) imply that for the triplet $(\tilde{a}, \tilde{A}, \tilde{\Pi})$ of Lévy characteristics for $\tilde{\mu}$ the following equalities hold true:

$$\tilde{a} = a + \int_{\{|y|>1\}} \frac{y}{|y|}\Pi(\mathrm{d}y)$$

$$\tilde{A} = \frac{1}{2}A$$

and

$$\tilde{\Pi}(B) = \int_{R^d}\int_0^\infty 1_B(e^{-s}y)\mathrm{d}s\,\Pi(\mathrm{d}y), \quad B \in \mathscr{B}(R_0^d).$$

Vice versa, if

$$\tilde{\Pi}(B) = \int_{S^{d-1}} \lambda(\mathrm{d}\xi)\int_0^\infty 1_B(r\xi)\frac{k_\xi(r)}{r}\mathrm{d}r, \quad B \in \mathscr{B}(R_0),$$

then

$$a = \tilde{a} - \int_{\{|y|>1\}} \frac{y}{|y|}\Pi(\mathrm{d}y),$$

$$A = 2\tilde{A} \tag{5.6}$$

and

$$\Pi(B) = -\int_{S^{d-1}} \lambda(\mathrm{d}\xi)\int_0^\infty 1_B(r\xi)\mathrm{d}k_\xi(r).$$

The process $\{Z_t, t \geq 0\}$, is called the background driving Lévy process (BDLP for short).

Definition 5.1 The subclass of the Ornstein–Uhlenbeck type processes, obtained by the correspondence (5.5) with the Student t-distribution $\tilde{\mu}$, is called the class of the Ornstein–Uhlenbeck type Student processes (Student OU-type processes for short).

Definition 5.2 The subclass of the Ornstein–Uhlenbeck type processes, obtained by the correspondence (5.5) with the noncentral Student t-distributions $\tilde{\mu}$, satisfying self-decomposability condition (iii) of Proposition 4.6, is called the class of the non-central Ornstein–Uhlenbeck type Student processes (noncentral Student OU-type processes for short).

We shall describe the BGDP, generating the Student OU-type processes.

Proposition 5.3 (i) *The Student OU-type processes are generated by the BDLP* $Z = \{Z_t, t \geq 0\}$ *with the triplets of Lévy characteristics* $(\gamma_0, 0, \Pi_0)$, *where*

$$\gamma_0 = \int\limits_{\{|x|\leq 1\}} x\pi_0(x)\mathrm{d}x + \alpha, \quad \alpha \in R^d,$$

$$\Pi_0(B) = \int\limits_B \pi_0(x)\mathrm{d}x, \quad B \in \mathscr{B}(R_0^d),$$

$$\pi_0(x) = -\frac{\mathrm{d}}{\mathrm{d}r}\left(r^d l_0(r\xi)\right)|_{r\xi=x}$$

and

$$l_0(x) = \frac{\nu 2^{\frac{d}{4}+1}\left(\langle x\Sigma^{-1}, x\rangle\right)^{-\frac{d}{4}}}{\sqrt{|\Sigma|}(2\pi)^{\frac{d}{2}}} \int\limits_0^\infty u^{\frac{d}{4}} K_{\frac{d}{2}}\left((2t\langle x\Sigma^{-1}, x\rangle)^{\frac{1}{2}}\right) g_{\frac{\nu}{2}}(2\nu t)\mathrm{d}t,$$

$\nu > 0$, Σ *is a symmetric positive definite* $d \times d$ *matrix.*

(ii) *The Student OU-type process* X, *generated by the BDLP* Z *with the triplet of Lévy characteristics* $(\gamma_0, 0, \Pi_0)$ *and* $\mathcal{L}(X_0) = T_d(\nu, \Sigma, \alpha)$ *is strictly stationary Markov process.*

Proof

(i) Follows directly from the Definition 5.1, the above stated properties of BGDP and the Proposition 4.5.
(ii) It is well-known property of time homogeneous Markov processes. □

Proposition 5.4

(i) *The noncentral Student OU-type processes are generated by the BDLP* $Z = \{Z_t, t \geq 0\}$ *with the triples of Lévy characteristics* $(\gamma_a, 0, \Pi_a)$, *where*

$$\gamma_a = \int\limits_{\{|x|\leq 1\}} x\pi_a(x)\mathrm{d}x + \alpha, \quad \alpha, a \in R^d,$$

$$\Pi_a(B) = \int\limits_B \pi_a(x)dx, \quad B \in \mathscr{B}(R_0^d),$$

$$\pi_a(x) = -\frac{\mathrm{d}}{\mathrm{d}r}\left(r^d l_a(r\xi)\right)|_{r\xi=x}$$

and

$$l_a(x) = \frac{2\nu \exp\left\{\langle a\Sigma^{-1}, x\rangle\right\}}{\sqrt{|\Sigma|}(2\pi)^{\frac{d}{2}}\left(\langle x\Sigma^{-1}, x\rangle\right)^{\frac{d}{4}}} \int_0^\infty \left(\langle a\Sigma^{-1}, a\rangle + 2t\right)^{\frac{d}{4}}$$
$$\times K_{\frac{d}{2}}\left(\left((\langle a\Sigma^{-1}, a\rangle + 2t)\langle x\Sigma^{-1}, x\rangle\right)^{\frac{1}{2}}\right) g_{\frac{\nu}{2}}(2\nu t).$$

(ii) *The noncentral Student OU-type process $X^{(a)}$, generated by the BDLP Z with the triplet of Lévy characteristics $(\gamma_a, 0, \Pi_a)$ and $\mathcal{L}(X_0) = T_d(\nu, \Sigma, \alpha, a)$ is strictly stationary Markov process.*

Proof

(i) Follows directly from the Definition 5.2, the above stated properties of BDLP and the Proposition 4.5.
(ii) It is well-known property of time homogeneous Markov processes. □

References

1. Sato, K.: Additive processes and stochastic integrals. Illinois J. Math **50**(4), 825–859 (2006)
2. Sato, K.: Lévy Processes and Infinitely Divisible Distributions. Cambridge University Press, Cambridge (1999)
3. Gravereaux, J.B.: Probabilité de Lévy sur R^d et équations différentielles stochastiques linéaires. In: Séminaire de Probabilités, Publications des Séminares de Matheématiques, Université de Rennes I, 1–42 (1982)
4. Jurek, Z.J., Verwaat, W.: An integral representation for self-decomposable Banach space valued random variables. Z. Wahrscheinlichkeitstheorie verw. geb., **52**, 247–262 (1983)
5. Sato, K., Yamazato, M.: Stationary processes of Ornstein-Uhlenbeck type. In: Probability Theory and Mathematical statistics (Tbilisi, 1982), Lecture Notes in Mathematics, vol. 1021, pp. 541–551. Springer, Berlin (1983)
6. Wolfe, S.J.: On a continuous analogue of the stochastic difference equation $X_n = \rho X_{n-1} + B_n$. Stoch. Process. Appl. **12**(3), 301–312 (1982)

Chapter 6
Student Diffusion Processes

6.1 H-Diffusions

We shall consider the regular positive recurrent diffusion processes $X = \{X_t, t \geq 0\}$ on an open interval $(l, r) \subseteq R^1$ with the inaccessible end points and predetermined one-dimensional distributions (for used terminology see, e.g., [1, 2]).

Let $\tau_a = \inf\{t > 0 : X_t = a\}$, $a \in (l, r)$, and $s(x)$, $s \in (l, r)$ be the scale function for the process X, i.e. a strictly increasing continuous function such that for all $l < a \leq x \leq b < r$

$$\mathrm{P}^x\{\tau_a < \tau_b\} = \frac{s(b) - s(x)}{s(b) - s(a)},$$

where P^x denotes the underlying probability measure of the process given $X_0 = x$.

Let m be the speed measure for the process X, characterized by the properties that $m(I) > 0$ for every non-empty subinterval I of (l, r) and for $l < a < x < b < r$

$$\mathrm{E}^x(\tau_a \wedge \tau_b) = \int\limits_{(a,b)} g_{s(a),s(b)}\,(s(x), s(y))\, m(dy)$$

where

$$g_{a,b}(u, v) = \begin{cases} \dfrac{(b-u)(v-a)}{b-a}, & \text{if } \ v \leq u, \\ \dfrac{(u-a)(b-v)}{b-a}, & \text{if } \ u \leq v, \end{cases}$$

and the expectation E^x is taken with respect to the measure P^x.

It is known (see [1–4]) that if $s(x) \to +\infty$, as $x \uparrow r$, $s(x) \to -\infty$, as $x \downarrow l$, and

$$|m| := m\left((l, r)\right) < \infty,$$

B. Grigelionis, *Student's* t*-Distribution and Related Stochastic Processes*, SpringerBriefs in Statistics, DOI: 10.1007/978-3-642-31146-8_6,

then the diffusion X is positive recurrent with the inaccessible end points. Moreover, if

$$\mathcal{L}(X_0) = \frac{m}{|m|},$$

the process X will be strictly stationary and ergodic.

Let $\mathcal{G}(l,r)$ be a class of strictly positive differentiable functions $g(x), x \in (l,r)$, such that for each $x \in (l,r)$ there exists $\varepsilon > 0$, $(x-\varepsilon, x+\varepsilon) \subset (l,r)$, satisfying

$$\int_{x-\varepsilon}^{x+\varepsilon} |g'(v)|dv < \infty,$$

for some $x_0 \in (l,r)$, as $x \uparrow r$,

$$G(x) := \int_{x_0}^{x} g(v)dv \to +\infty,$$

and, as $x \downarrow l$, $G(x) \to -\infty$.

Let $h(x), x \in (l,r)$ be a strictly positive measurable function such that

$$\int_{l}^{r} h(x)dx = 1. \tag{6.1}$$

Write $H(\mathrm{d}x) = h(x)\mathrm{d}x$,

$$a(x) = -\frac{1}{2}\frac{g'(x)}{h(x)g^2(x)}, \quad x \in (l,r), \tag{6.2}$$

and

$$\sigma^2(x) = (h(x)g(x))^{-1}, \quad x \in (l,r). \tag{6.3}$$

Theorem 6.1 [5] *For each $g \in \mathcal{G}(l,r)$ and h, satisfying* (6.1), *there exists the unique weak solution for the stochastic differential equation*

$$\begin{cases} dX_t = a(X_t)dt + \sigma(X_t)dB_t, & t > 0 \\ \mathcal{L}(X_0) = H, \end{cases}$$

which is a regular positive recurrent diffusion with the scale function

$$s(x) = \int_{x_0}^{x} \frac{g(v)}{g(x_0)} dv, \quad x \in (l, r),$$

and the speed measure $m = g(x_0)H$.

Here and below $B = \{B_t, t \geq 0\}$ *is the standard univariate Brownian motion.*

The solution is a strictly stationary process with the one dimensional distribution H, called the H-diffusion (see [6]). The functions g and h are intrinsic characteristics of the H-diffusions, in terms of which their properties should be formulated.

Example 6.2 Let $(l, r) = (0, 1)$,

$$\begin{aligned}
h(x) &= Cx^{\beta_1 - 1}(1-x)^{\beta_2 - 1} e^{\lambda x}, \quad x \in (0, 1), \\
g(x) &= \frac{1}{C\sigma^2}\left[x^{\alpha_1 + \beta_1 - 1}(1-x)^{\alpha_2 + \beta_2 - 1} e^{(\chi + \lambda)x}\right]^{-1}, \\
x &\in (0, 1), \quad \alpha_1, \alpha_2, \lambda, \chi \in R^1, \quad \sigma^2 > 0, \quad \beta_1 > 0, \quad \beta_2 > 0.
\end{aligned}$$

Here and below C is the norming constant. It is easy to check that $g \in \mathscr{G}(0, 1)$ if and only if $\alpha_1 + \beta_1 \geq 2$ and $\alpha_2 + \beta_2 \geq 2$.

In this case

$$\begin{aligned}
a(x) = \frac{\sigma^2}{2}\Big[&(\alpha_1 + \beta_1 - 1)x^{\alpha_1 - 1}(1-x)^{\alpha_2} - (\alpha_2 + \beta_2 - 1) \\
&\times x^{\alpha_1}(1-x)^{\alpha_2 - 1}(\lambda + \mu)x^{\alpha_1}(1-x)^{\alpha_2}\Big] e^{\chi x}, \quad x \in (0, 1),
\end{aligned}$$

and

$$\sigma^2(x) = \sigma^2 x^{\alpha_1}(1-x)^{\alpha_2} e^{\chi x}, \quad x \in (0, 1).$$

Taking $\alpha_1 = \alpha_2 = 1$, $\chi = 0$, we have the Wright–Fisher gene frequency model with mutation and selection in the population genetics (see, e.g., [1, 7]).

Example 6.3 Let $(l, r) = (0, \infty)$,

$$\begin{aligned}
h(x) &= Cx^{\lambda - 1}\exp\left\{-\left(\chi x^{-\beta_1} + \psi x^{\beta_2}\right)\right\}, \quad x > 0 \\
g(x) &= \frac{1}{C\sigma^2} x^{-(\lambda + \gamma) + 1}\exp\left\{\chi x^{-\beta_1} + \psi x^{\beta_2}\right\}, \quad x > 0
\end{aligned}$$

where $\sigma^2 > 0$, $\beta_1 > 0$, $\beta_2 > 0$ and either

(i) $\lambda, \gamma \in R^1$, $\chi > 0$, $\psi > 0$, or
(ii) $\chi = 0$, $\lambda > 0$, $\psi > 0$, $\lambda + \gamma > 2$, or
(iii) $\psi = 0$, $\lambda < 0$, $\chi > 0$, $\lambda + \gamma < 2$.

In all these cases $g \in \mathcal{G}(0, \infty)$,

$$a(x) = \frac{\sigma^2}{2}\left[(\gamma + \lambda - 1)x^{\gamma-1} + \chi\beta_1 x^{\gamma-\beta_1-1} - \psi\beta_2 x^{\gamma+\beta_2-1}\right], \quad x > 0$$

and

$$\sigma^2(x) = \sigma^2 x^\gamma, \quad x > 0.$$

If $\gamma = 2$, $\beta_2 = 1$, $\chi = 0$, $\lambda > 0$, we have that

$$a(x) = \frac{\sigma^2}{2}(\lambda + 1)x - \psi x^2, \quad x > 0,$$

$$\sigma^2(x) = \sigma^2 x^2$$

and

$$h(x) = Cx^{\lambda-1}e^{-\psi x},$$

giving us a diffusion version of the Pearl-Verhulst logistic population growth model (see [1]). This class of diffusions also contains the Cox–Ingersoll–Ross model for short interest rates in bond markets and its generalizations (see, e.g., [4, 8]).

Example 6.4 Let $(l, r) = (-\infty, +\infty)$,

$$h(x) = C\left(1 + \left(\frac{x-\alpha}{\delta}\right)^2\right)^\gamma \exp\left\{-\varkappa \arctan\left(\frac{x-\alpha}{\delta}\right)\right\}, \quad x \in R^1,$$

$$g(x) = \frac{\exp\left\{\varkappa \arctan\left(\frac{x-\alpha}{\delta}\right)\right\}}{C\sigma^2\left(1 + \left(\frac{x-\alpha}{\delta}\right)^2\right)^{\lambda+\gamma}}, \quad x \in R^1 \quad \alpha, \lambda, \varkappa \in R^1, \quad \lambda < -\frac{1}{2},$$

$$\lambda + \gamma \leq \frac{1}{2}, \quad \delta > 0, \quad \sigma^2 > 0.$$

In this case $g \in \mathcal{G}(-\infty, +\infty)$,

$$\mu(x) = \frac{\sigma^2}{\delta}\left(1 + \left(\frac{x-\alpha}{\delta}\right)^2\right)^{\gamma-1}\left[(\lambda + \gamma)\left(\frac{x-\alpha}{\delta}\right) - \frac{\varkappa}{2}\right], \quad x \in R^1,$$

$$\sigma^2(x) = \sigma^2\left(1 + \left(\frac{x-\alpha}{\delta}\right)^2\right)^\gamma, \quad x \in R^1.$$

Taking $\gamma = 1$, we have the Johannesma diffusion model for the stochastic activity of neurons (see [9–11]) and one of the Föllmer–Schweizer models for stock returns (see [12], also [13]). The stationary distribution is the skew Student's t-distribution with the skewness coefficient $\varkappa$. If $\varkappa = 0$, we arrive to the univariate Student's t-distribution.

6.2 Student Diffusions

Definition 6.5 An H-diffusion process X on R^1 is called a Student diffusion if $H = T_1(\nu, \sigma^2, \alpha)$, $\nu > 0, \sigma^2 > 0, \alpha \in R^1$.

From Theorem 6.1 it follows that for each $g \in \mathcal{G}(-\infty, \infty)$ there exists a Student diffusion. For example, taking $\lambda = -\frac{\nu+1}{2}$, $\gamma = 1$, $\varkappa = 0$, $\sigma^2 = \theta$, we find from Example 6.4 that the unique weak solution for the stochastic differential equation

$$\begin{cases} \mathrm{d}X_t = -\dfrac{\theta(\nu-1)}{2}\left(\dfrac{X_t-\alpha}{\delta}\right)\mathrm{d}t + \sqrt{\theta\left(1+\left(\dfrac{X_t-\alpha}{\delta}\right)^2\right)}\mathrm{d}B_t, \quad \theta > 0, \\ \mathcal{L}(X_0) = T_1(\nu, \delta^2\nu^{-1}, \alpha) \end{cases}$$

is a Student diffusion.

Example 6.6 [8] The function $g(x) \equiv \sigma^{-2} > 0$, $x \in R^1$, belongs to $\mathcal{G}(-\infty, \infty)$. Thus for any strictly positive pdf $h(x)$, $x \in R^1$, the unique weak solution for the stochastic differential equation

$$\begin{cases} \mathrm{d}X_t = \left(\sigma^2 h(X_t)\right)^{-\frac{1}{2}} \mathrm{d}B_t, \quad t > 0 \\ \mathcal{L}(X_0) = H, \end{cases}$$

is an H-diffusion.

If $\nu > 1$, as the unique weak solution for the stochastic differential equation

$$\begin{cases} \mathrm{d}X_t = -\theta\dfrac{X_t-\alpha}{\delta}\mathrm{d}t + \sqrt{\dfrac{2\theta\delta^2}{\nu-1}\left(1+\left(\dfrac{X_t-\alpha}{\delta}\right)^2\right)}\mathrm{d}B_t, \quad t > 0, \\ \mathcal{L}(X_0) = T_1(\nu, \delta^2\nu^{-1}, \alpha), \end{cases}$$

the Student diffusion is a member of the family of Kolmogorov–Pearson diffusions (see [14, 15]).

Now let us consider a Student diffusion $X = \{X_t, t \geq 0\}$, corresponding to the function $g \in \mathcal{G}(-\infty, \infty)$, and discuss the domain-of-attraction problem for the maximum values

$$M_T = \max_{0 \le t \le T} X_t, \quad T > 0,$$

using linear normalization.

We shall see that the problem for H-diffusion reduces to the classical extreme value theory and the criteria are expressed in the terms of functions g independently of the marginal distribution H.

Definition 6.7 We say that an H-diffusion $X = \{X_t, t \ge 0\}$ belongs to the maximum domain of attraction of the nondegenerate distribution Q ($X \in MDA_l(Q)$ for short) if there exist constants $a_T > 0$ and $b_T \in R^1$ such that, as $T \to \infty$,

$$\mathscr{L}(a_T(M_T - b_T)) \Rightarrow Q.$$

Define γ_T from the equality $G(\gamma_T) = T$.

Theorem 6.8 [6] *Let an* H*-diffusion X corresponds to the function $g \in \mathscr{G}(l, r)$. The following criteria hold true:*

(i) $X \in MDA_l(\Lambda)$ if and only if there exists a function $b(x) > 0$, $x \in (x_0, r)$, such that, for each $x \in R^1$,

$$\lim_{y \uparrow r} \frac{G(y)}{G(y + b(y)x)} = e^{-x};$$

(ii) $X \in MDA_l(\Phi_\gamma)$ if and only if $r = \infty$ and, for each $x > 0$,

$$\lim_{y \uparrow \infty} \frac{G(y)}{G(xy)} = x^{-\gamma}, \quad \gamma > 0;$$

(iii) $X \in MDA_l(\Psi_\gamma)$ if and only if $r < \infty$ and, for each $x > 0$

$$\lim_{y \downarrow 0} \frac{G(r - y)}{G(r - xy)} = x^{\gamma}, \quad \gamma > 0.$$

Moreover, in the case (i)

$$\int_{x_0}^{r} (G(v))^{-1} \, dv < \infty$$

and we can take

$$b(x) = G(x) \int_x^r (G(v))^{-1}\, dv,$$

$$a_T \sim \frac{1}{T}\left(\int_{\gamma_T}^r (G(v))^{-1}\, dv\right)^{-1},$$

$$b_T = \gamma_T + \chi_T,$$

where χ_T *are any constants such that* $a_T \chi_T \to 0$, *as* $T \to \infty$.

In the case (ii)

$$a_T \sim \gamma_T^{-1}, \quad b_T = 0$$

and in the case (iii)

$$a_T \sim (r - \gamma_T)^{-1}, \quad b_T = r.$$

Proof Under the assumptions of Theorem from Davis [16] (see also [17, 18]) we have that for any constants $u_T \uparrow \infty$, as $T \to \infty$.

$$\lim_{T\to\infty} \left| \mathrm{P}\{M_T \le u_T\} - F^T(u_T) \right| = 0,$$

where

$$F(x) = e^{-(G(x))^{-1}}, \quad x \in (l, r).$$

Let

$$\hat{F}(x) = \begin{cases} 0, & \text{for} \quad x < \hat{x}_0, \\ 1 - (G(x))^{-1}\, 1_{(\hat{x}_0, r)}, & \text{for} \quad x \ge \hat{x}_0, \end{cases}$$

where $G(\hat{x}_0) = 1$.

Because $1 - F(x) \sim 1 - \hat{F}(x)$, as $x \uparrow r$, the statement of Theorem 6.8, using the principle of equivalent tails, now follows from the classical extreme value theory (see, e.g., [19, 20]). □

Because, for $x \in (\hat{x}_0, r)$,

$$\hat{f}(x) := \hat{F}'(x) = \frac{g(x)}{2G^2(x)}$$

and

$$\hat{f}'(x) = \frac{1}{2}\frac{g'(x)}{G^2(x)} - \frac{g^2(x)}{G^3(x)},$$

we shall have the following analogue of classical von Mises theorem (see, [19–22]).

Theorem 6.9 *[21] Let an* H-*diffusion X correspond to the function* $g \in \mathcal{G}(l, r)$.
The following sufficient conditions are valid:

(i) if

$$\lim_{x \uparrow r} \frac{g'(x)G(x)}{g^2(x)} = 1,$$

then $X \in MDA_l(\Lambda)$;

(ii) if $r = \infty$ *and*

$$\lim_{x \uparrow \infty} \frac{xg(x)}{G(x)} = \gamma > 0,$$

then $X \in MDA_l(\Phi_\gamma)$;

(iii) if $r < \infty$ *and*

$$\lim_{x \uparrow r} \frac{(r-x)g(x)}{G(x)} = \gamma > 0,$$

then $X \in MDA_l(\Psi_\gamma)$.

Now the following Propositions are obvious.

Proposition 6.10 *Let a Student diffusion X correspond to the function* $g \in \mathcal{G}(-\infty, \infty)$.
There are two possibilities:

(1) $X \in MDA_l(\Lambda)$ *if and only if there exists a function* $b(x) > 0$, $x \in (x_0, \infty)$, *such that, for each* $x \in R^1$,

$$\lim_{y \uparrow \infty} \frac{G(y)}{G(y + b(y)x)} = e^{-x},$$

and

(2) $X \in MDA_l(\Phi_\gamma)$ *if and only if, for each* $x > 0$,

$$\lim_{y \uparrow \infty} \frac{G(y)}{G(xy)} = x^{-\gamma}, \quad \gamma > 0.$$

In the case (1) we can take

$$b(x) = G(x) \left(\int_{\infty_T}^{\infty} (G(v))^{-1} \, dv \right)^{-1},$$

and the norming constants

$$a_T \sim \frac{1}{T} \left(\int_{\gamma_T}^{\infty} (G(v))^{-1} \, dv \right)^{-1},$$
$$b_T = \gamma_T + \chi_T,$$

where χ_T are any constants such that $a_T \chi_T \to 0$, as $T \to \infty$.
In the case (2) the norming constants are $a_T \sim \gamma_T^{-1}$, $b_T = 0$.

Proposition 6.11 *Let a Student diffusion X correspond to the function $g \in \mathcal{G}(-\infty, \infty)$.*
Then, if

$$\lim_{x \uparrow \infty} \frac{g'(x)G(x)}{g^2(x)} = 1,$$

$X \in MDA_l(\Lambda)$,
and, if

$$\lim_{x \uparrow \infty} \frac{xg(x)}{G(x)} = \gamma > 0,$$

$X \in MDA_l(\Psi_\gamma)$.

Example 6.12 (continued Example 6.2) Let $\alpha_1 + \beta_1 > 2$. Using Theorem 6.9 (iii), because

$$\lim_{x \uparrow 1} \frac{(1-x)g(x)}{G(x)} = \alpha_1 + \beta_1 - 2,$$

$X \in MDA_l(\Psi_{\alpha_1+\beta_1-2})$.

Example 6.13 (continued Example 6.3) In the both cases (i) and (ii)

$$\lim_{x \to \infty} \frac{g'(x)G(x)}{g^2(x)} = 1,$$

implying by Theorem 6.9 (i) that $X \in MDA_l(\Lambda)$.

In the case (iii), assuming that $\lambda + \gamma < 2$, we have that

$$\lim_{x \uparrow \infty} \frac{xg(x)}{G(x)} = 2 - \lambda - \gamma,$$

implying by Theorem 6.9 (ii) that $X \in MDA_l(\Phi_{2-\lambda-\gamma})$.

Example 6.14 (continued Example 6.4) Assuming that $\lambda + \gamma < \frac{1}{2}$, we have that

$$\lim_{x \uparrow \infty} \frac{xg(x)}{G(x)} = 1 - 2(\lambda + \gamma),$$

implying by Theorem 6.9 (ii) that $X \in MDA_l(\Phi_{1-2(\lambda+\gamma)})$.

Example 6.15 (continued Example 6.6) Taking $x_0 = 0$, we find that $G(x) = \sigma^2 x$, $x \in R^1$, $\gamma_T = \sigma^{-2} T$ and

$$\frac{xg(x)}{G(x)} \equiv 1.$$

Thus, $X \in MDA_l(\Phi_1)$ and, as $T \to \infty$,

$$\mathscr{L}\left(\frac{\sigma^2}{T} M_T\right) \Rightarrow \Phi_1.$$

6.3 Point Measures of ε-Upcrossings for Student Diffusions

Let $\varepsilon > 0$ be fixed. The process $X = \{X_t, t \geq 0\}$ is said to have an ε-upcrossing of the level u at t_0 if $X(t) < u$, for $t \in (t_0 - \varepsilon, t_0)$, and $X(t_0) = u$. Let $T > 0$ and $B \in \mathscr{B}((0, 1])$. Then

$$N_T(B) = \sharp\{\varepsilon - \text{crossings of } u_T \text{ by } X \text{ on the set } TB\}$$

is called the time normalized point measure of ε-upcrossings of the level u_T by X.

The following statement is slightly weakened but essentially simplified version of the Borkovec and Klüppelberg result in [8] (for used terminology see, e.g., [23]).

Theorem 6.16 [24] *Let an* H-*diffusion X correspond to the function $g \in \mathscr{G}(l, r)$, pdf h is continuous and there exists a constant K such that, for all $x \in (l, r)$,*

$$\frac{h(x) G^2(x) \log(|G(x)| + 1)}{g(x)} \leq K. \tag{6.4}$$

If $u_T \uparrow r$, as $T \to \infty$, and

$$\lim_{T\to\infty} T^{-1} G(u_T) = (2\tau)^{-1}, \quad \tau > 0, \tag{6.5}$$

then the point measure N_T converges vaguely to the homogeneous Poisson point measure on $\mathscr{B}((0,1))$ with the intensity τ, as $T \to \infty$.

Example 6.17 Let $(l, r) = (-\infty, \infty)$, $x_0 = 0$, $h(x)$, $x \in R^1$, be an arbitrary strictly positive continuous pdf, $g(x) \equiv \sigma^{-2} > 0$.

If there exists a constant K such that, for all $x \in R^1$

$$x^2 \log(|x| + 1)\, h(x) \le K, \tag{6.6}$$

then the statement of Theorem 6.16 holds true with $\tau = \frac{\sigma^2}{2}$ and $u_T = T$.

Because for the skew Student's t-distribution (see Example 6.4 and [13])

$$h(x) = C_{\nu,\delta,\varkappa}\left(1 + \left(\frac{x-\alpha}{\delta}\right)^2\right)^{-\frac{\nu+1}{2}} \exp\left\{-\varkappa \arctan\left(\frac{x-\alpha}{\delta}\right)\right\}, \quad x \in R^1, \tag{6.7}$$

where

$$C_{\nu,\delta,\varkappa} = \frac{\Gamma\left(\frac{\nu+1}{2}\right)}{\delta\sqrt{\pi}\,\Gamma\left(\frac{\nu}{2}\right)} \prod_{k=0}^{\infty}\left[1 + \frac{\varkappa^2}{(\nu+1+2k)^2}\right]^{-1},$$

we have that, as $|x| \to \infty$,

$$h(x) \sim C_{\nu,\delta,\varkappa}\delta^{\nu+1}|x|^{-(\nu+1)}. \tag{6.8}$$

In this case the assumption (6.4) is satisfied if and only if $\nu > 1$.

Example 6.18 Let X be a skew Student diffusion corresponding to the function

$$g(x) = \frac{\exp\left\{\varkappa \arctan\left(\frac{x-\alpha}{\delta}\right)\right\}}{C_{\nu,\delta,\varkappa}\left(1 + \left(\frac{x-\alpha}{\delta}\right)^2\right)^{-\frac{\nu+1}{2}+\gamma}}, \quad x \in R^1, \quad \alpha, \varkappa \in R^1, \quad \gamma \le 1 + \frac{\nu}{2}.$$

Having in mind (6.8), because, as $|x| \to \infty$,

$$g(x) \sim \frac{|x|^{\nu+1-2\gamma}}{C_{\nu,\delta,\varkappa}\delta^{\nu+1-2\gamma}}$$

and, using l'Hospital's rule,

$$G(x) \sim \frac{|x|^{\nu+2-2\gamma}}{C_{\nu,\delta,\varkappa}\delta^{\nu+1-2\gamma}(\nu+2-2\gamma)},$$

we find that the assumption (6.6) is satisfied if and only if $1 < \gamma \leq 1 + \frac{\nu}{2}$.

If $1 < \gamma < 1 + \frac{\nu}{2}$, taking

$$u_T = \left(\frac{T}{2C_{\nu,\delta,\varkappa}\delta^{\nu+1-2\gamma}(\nu+2-2\gamma)}\right)^{\frac{1}{\nu+2-2\gamma}},$$

then the point measure N_T, as $T \to \infty$, converge vaguely to the Poisson measure with the intensity 1.

Example 6.19 (continued Example 6.3) In the case (i), using l'Hospital's rule, we have that, as $x \to \infty$,

$$G(x) \sim (\psi\beta_1)^{-1}x^{1-\beta_1}g(x) \tag{6.9}$$

and, as $x \to 0$,

$$G(x) \sim -(\chi\beta_2)^{-1}x^{1+\beta_2}g(x). \tag{6.10}$$

Thus, the assumption (6.4) is satisfied if and only if

$$2 - 2\beta_1 < \gamma < 2 + 2\beta_2$$

and (6.5) holds with $\tau = 1$ and

$$\begin{aligned} u_T = {} & \left(\frac{1}{\psi}\log T\right)^{\frac{1}{\beta_1}} + \frac{1}{\beta_1\psi}\left(\frac{1}{\psi}\log T\right)^{\frac{1}{\beta_1}-1} \\ & \times \left[\frac{\beta_1+\gamma+\lambda-2}{\beta_1}\log\left(\frac{1}{\psi}\log T\right) + \log\left(\beta_1\psi C\frac{\sigma^2}{2}\right)\right] \end{aligned} \tag{6.11}$$

Here we used formulas for asymptotic solutions of equations like $G(u_T) = T$ from [19], Table 3.4.4.

In the case (ii) we analogously find that, as $x \to \infty$, (6.10) holds, and, as $x \to 0$,

$$G(x) \sim \frac{x}{2-(\lambda+\gamma)}g(x), \tag{6.12}$$

implying that the assumption (6.4) is satisfied if and only if

$$2 < \gamma < 2\beta_2 + 2.$$

The equality (6.5) holds with $\tau = 1$ and u_T, defind by (6.11).

Finally, in the case (iii), as $x \to \infty$, it holds (6.12) and, as $x \to 0$, it holds (6.10), implying that the assumption (6.4) is satisfied if and only if

$$2 - 2\beta_1 < \gamma < 2.$$

The equality (6.5) holds with $\tau = 1$ and

$$u_T = \left[(2 - \lambda - \gamma)(\frac{\sigma^2 T}{2}) \right]^{\frac{1}{2-\lambda-\gamma}}.$$

6.4 Kolmogorov–Pearson Diffusions

Definition 6.20 An H-diffusion $X = \{X_t, t \geq 0\}$ in the interval (l, r) is called the Kolmogorov–Pearson diffusion if it is a weak solution for the stochastic differential equation

$$\begin{cases} \mathrm{d}X_t = \theta A(X_t)\mathrm{d}t + \sqrt{\theta B(X_t)}\mathrm{d}B_t, \quad t > 0, \quad \theta > 0, \\ \mathscr{L}(X_0) = H, \end{cases} \tag{6.13}$$

where

$$A(x) = p_0 + p_1 x, \quad x \in (l, r),$$

and

$$B(x) = q_0 + q_1 x + q_2 x^2 > 0, \quad x \in (l, r).$$

This class of diffusions was described by Kolmogorov in 1931 (see [25]). Ergodic distributions of these diffusions are contained in the family of Pearson distributions, satisfying the Pearson equation:

$$\frac{h'(x)}{h(x)} = \frac{2A(x) - B'(x)}{B(x)}, \quad x \in (l, r). \tag{6.14}$$

Last years this class of diffusions attracted attention of statisticians as a flexible and statistically tractable stochastic processes (see, e.g., [13, 26–32]).

Let $L^2\left((l, r); H\right)$ be a Hilbert space of equivalency classes of measurable functions $f : (l, r) \to R^1$ such that

$$||f||_H^2 := \int_l^r f^2(x)h(x)\mathrm{d}x < \infty$$

and $C^2((l,r))$ be a class of twice differentiable functions $f:(l,r)\to R^1$.

The generator

$$L=\frac{\theta}{2}B(x)\frac{\mathrm{d}^2}{\mathrm{d}x^2}+\theta A(x)\frac{\mathrm{d}}{\mathrm{d}x}$$

of the Kolmogorov–Pearson diffusion X, satisfying (6.13), is a map

$$L:L^2((l,r);H)\cap C^2((l,r))\to L^2((l,r);H).$$

Let us recall the following classical results (see, e.g., [1, 33–35]).

Obviously, L maps polynomials to polynomials. If, for all $n=0,1,\ldots$,

$$\int_l^r x^{2n}h(x)\mathrm{d}x<\infty,$$

there exists an orthonormal system of polynomials $\{P_n(x), x\in(l,r), n=0,1,\ldots\}$ such that

$$LP_n(x)+\lambda_nP_n(x)=0,\quad x\in(l,r),$$

where

$$\lambda_n=-n\theta\left(p_1+\frac{q_2}{2}(n+1)\right),\quad n=0,1,\ldots,\tag{6.15}$$

showing that the spectrum of—L is discrete with the eigenvalues, given by (6.15), and the corresponding eigenfunctions $\{P_n(x), x\in(l,r), n=0,1,\ldots\}$, which under the additional assumption that

$$\lim_{x\to l-0}h(x)B(x)=\lim_{x\to r+0}h(x)B(x)=0\tag{6.16}$$

are given by the generalized Rodrigues formula:

$$P_n(x)=c_n\frac{\left[h(x)B^n(x)\right]^{(n)}}{h(x)},\quad x\in(l,r),\quad n=0,1,\ldots,\tag{6.17}$$

where

$$c_n^{-2}=\int_l^r\frac{\left(\left[h(x)B^n(x)\right]^{(n)}\right)^2}{h(x)}\mathrm{d}x.$$

If, for some integer N,

$$\int_{l}^{r} x^{2N} h(x)\mathrm{d}x < \infty, \tag{6.18}$$

but

$$\int_{l}^{r} |x|^{2N+1} h(x)\mathrm{d}x = \infty,$$

the spectrum of—L consists of the continuous part and the finite number of discrete eigenvalues

$$\lambda_n = -n\theta\left(p_1 + \frac{q_2}{2}(n+1)\right), \quad n = 0, 1, \ldots, N,$$

corresponding to the eigenfunctions $\{P_n(x), x \in (l, r), n = 0, 1, \ldots, N\}$, defined by the formula (6.17).

Let

$$h_j = \int_{l}^{r} x^j h(x)\mathrm{d}x, \quad j = 0, 1, 2, \ldots,$$

$$\Delta_n = \begin{vmatrix} 1 & h_1 & \ldots & h_n \\ h_1 & h_2 & \ldots & h_{n+1} \\ \ldots & \ldots & \ldots & \ldots \\ h_n & h_{n+1} & \ldots & h_{2n} \end{vmatrix}, \quad \Delta_0 = 1,$$

and

$$Q_n(x) = \begin{vmatrix} 1 & h_1 & \ldots & h_n \\ h_1 & h_2 & \ldots & h_{n+1} \\ \ldots & \ldots & \ldots & \ldots \\ h_{n-1} & h_n & \ldots & h_{2n-1} \\ 1 & x & \ldots & x^n \end{vmatrix}, \quad Q_0(x) \equiv 1.$$

Then

$$P_n(x) = \frac{Q_n(x)}{\sqrt{\Delta_{n-1}\Delta_n}}, \quad x \in (l, r), \quad n = 1, 2, \ldots.$$

If h is a pdf of the skew Student's t-distribution, given by (6.7), from Example 6.4 it follows that the corresponding H-diffusion is the Kolmogorov–Pearson diffusion with

$$A(x) = \frac{\theta}{\delta}\left[-\frac{\nu-1}{2}\left(\frac{x-\alpha}{\delta}\right) - \frac{\varkappa}{2}\right], \tag{6.19}$$

and

$$B(x) = \theta\left(1 + \left(\frac{x-\alpha}{\delta}\right)^2\right), \quad x \in R^1, \quad \alpha, \varkappa \in R^1, \quad \nu, \delta > 0. \tag{6.20}$$

In this case from (6.8) it follows that (6.16) is satisfied if and only if $\nu > 1$, and (6.18) holds true with the largest integer N satisfying $2N < \nu$ and denoted $N = \left\lfloor \frac{\nu}{2} \right\rfloor$. The discrete eigenvalues for the skew Student diffusion, defined by (6.19) and (6.20), are

$$\lambda_n = \frac{n\theta}{2\delta^2}(\nu - n), \quad n = 0, 1, \ldots, \left\lfloor \frac{\nu}{2} \right\rfloor.$$

The corresponding eigenfunctions are equal to

$$P_n(x) = c_n \frac{\left[h(x)\left(1 + \left(\frac{x-\alpha}{\delta}\right)^2\right)^n\right]^{(n)}}{h(x)}, \quad n = 0, 1, \ldots, \left\lfloor \frac{\nu}{2} \right\rfloor \tag{6.21}$$

If $\varkappa = 0$, h is the pdf of $T_1(\nu, \delta^2\nu^{-1}, \alpha)$. Following [30], polynomials (6.21) are called the Routh–Romanovsky polynomials (see [36, 37]).

If $\varkappa = \alpha = 0$, we have that, for $j < \nu$,

$$h_j^{(0)} := \begin{cases} \int\limits_{-\infty}^{\infty} x^j h(x)\mathrm{d}x = \dfrac{\delta^j}{\sqrt{\pi}\Gamma\left(\frac{\nu}{2}\right)}\Gamma\left(\dfrac{j}{2} + \dfrac{1}{2}\right)\Gamma\left(\dfrac{\nu}{2} - \dfrac{j}{2}\right), & \text{if } j \text{ is even}, \\ 0, \quad \text{if } j \text{ is odd}, \end{cases}$$

and, for $\varkappa = 0, \alpha \neq 0, j < \nu$,

$$h_j^{(\alpha)} := \int\limits_{-\infty}^{\infty} x^j h(x)\mathrm{d}x = \sum_{k=0}^{j} \binom{j}{k} h_k^{(0)} \alpha^{j-k}.$$

We refer the reader to [30] (see also [9, 15]) where a version of the Student diffusion was considered with

$$A(x) = \frac{-x + \alpha}{\delta},$$

$$B(x) = \frac{2\delta^2}{\nu - 1}\left(1 + \left(\frac{x-\alpha}{\delta}\right)^2\right), \quad \alpha \in R^1, \quad \nu > 1, \quad \delta > 0,$$
$$\lambda_n = \frac{\theta}{\nu - 1} n(\nu - n), \quad n = 0, 1, \ldots, \left\lfloor \frac{\nu}{2} \right\rfloor$$

and the Routh–Romanovsky polynomials as corresponding eigenfunctions. Most important that in this paper the continuous part of spectrum is described in terms of the hypergeometric functions, obtained the spectral representation of transition probability density of X and applied to the statistical inference of the model.

The skew Student diffusion is known as the Johannesma diffusion model for the stochastic activity of neurons (see [9–11]) and as one of the Föllmer–Schweizer models for stock returns (see [12, 13]).

Classification of the Kolmogorov–Pearson diffusions to six types is given in [14, 15]. The characteristics of these types are the following:

(1)

$$A(x) = -x + \alpha, \quad B(x) \equiv 2, \quad (l, r) = (-\infty, \infty),$$
$$h(x) = \frac{1}{\sqrt{2\pi}} e^{-\frac{1}{2}(x-\alpha)^2}, \quad x, \alpha \in R^1,$$
$$\lambda_n = n^2\theta, \quad n = 0, 1, \ldots$$

$\{P_n(x), x \in R^1, n = 0, 1, \ldots\}$ are the Hermite polynomials;

(2)

$$A(x) = -x + \alpha, \quad B(x) = 2x, \quad (l, r) = (0, \infty), \quad \alpha > 1,$$
$$h(x) = \frac{x^{\alpha-1}e^{-x}}{\Gamma(\alpha)}, \quad x > 0,$$
$$\lambda_n = n\theta, \quad n = 0, 1, \ldots,$$

$\{P_n(x), x > 0, n = 0, 1, \ldots\}$ are the Laguerre polynomials;

(3)

$$A(x) = -x + \alpha, \quad B(x) = 2ax^2, \quad (l, r) = (0, \infty), \quad a > 0, \quad \alpha > 0,$$
$$h(x) = C_{a^{-1}+1,1,\frac{\alpha}{a}} \left(1 + x^2\right)^{-\frac{1}{2a}-1} \exp\left\{-\frac{\alpha}{a} \arctan(x - \alpha)\right\}, \quad x > 0$$
$$\lambda_n = n\theta\left(1 - a(n+1)\right), \quad n = 0, 1, \ldots, \left\lfloor \frac{1}{2} + \frac{1}{2a} \right\rfloor,$$

$\{P_n(x), x > 0, n = 0, 1, \ldots, \lfloor \frac{1}{2} + \frac{1}{2a} \rfloor\}$ are the Routh-Romanovsky polynomials;

(4)

$$A(x) = -x + \alpha, \quad B(x) = 2ax^2, \quad (l, r) = (0, \infty), \quad a > 0, \quad \alpha > 0,$$
$$h(x) = \frac{\left(\frac{\alpha}{a}\right)^{\frac{1}{a}+1}}{\Gamma\left(\frac{1}{a}+1\right)} x^{-\frac{1}{a}-2} \exp\left\{-\frac{\alpha}{ax}\right\}, \quad x > 0,$$
$$\lambda_n = n\theta\,(1 - a(n+1)), \quad n = 0, 1, \ldots, \left\lfloor \frac{1}{2} + \frac{1}{2a} \right\rfloor,$$

$\left\{P_n(x), x > 0, n = 0, 1, \ldots, \left\lfloor \frac{1}{2} + \frac{1}{2a} \right\rfloor\right\}$ are the Bessel polynomials;

(5)

$$A(x) = -x + \alpha, \quad B(x) = 2ax(x+1), \quad (l, r) = (0, \infty), \quad \alpha \geq a > 0,$$
$$h(x) = \frac{1}{B\left(\frac{\alpha}{a}, \frac{1}{a}+1\right)} x^{\frac{\alpha}{a}-1} (1+x)^{-\frac{\alpha+1}{a}-1}, \quad x > 0,$$
$$\lambda_n = n\theta\,(1 - a(n+1)), \quad n = 0, 1, \ldots, \left\lfloor \frac{1}{2} + \frac{1}{2a} \right\rfloor,$$

$\left\{P_n(x), x > 0, n = 0, 1, \ldots, \left\lfloor \frac{1}{2} + \frac{1}{2a} \right\rfloor\right\}$ are the Fisher–Snedocor polynomials;

(6)

$$A(x) = -x + \alpha, \quad B(x) = 2ax(x-1), \quad (l, r) = (0, 1), \quad -1 < a < 0,$$
$$1 + a \leq \alpha \leq -a,$$
$$h(x) = \frac{1}{B\left(-\frac{\alpha}{a}, -\frac{1-\alpha}{a}\right)} x^{-\frac{\alpha}{a}-1} (1-x)^{-\frac{\alpha+1}{a}-1}, \quad 0 < x < 1,$$
$$\lambda_n = n\theta\,(1 - 2a(n+1)), \quad n = 0, 1, \ldots,$$

$\{P_n(x), x \in (0, 1), n = 0, 1, \ldots\}$ are Jacobi polynomials.

In the above formulas $B(z_1, z_2)$ means the Euler's beta function.

References

1. Karlin, S., Taylor, H.M.: A Second Course in Stochastic processes. Academic Press, New York (1981)
2. Mandl, P.: Analytical Treatment of One-Dimensional Markov Processes. Springer-Verlag, New York (1968)
3. Itô, K., McKean, H.: Diffusion Processes and Their Sample Paths. Springer, New York (1974)
4. Karatzas, I., Shreve, S.E.: Brownian Motion and Stochastic Calculus. Springer-Verlag, New-York (1998)
5. Engelbert, H.J., Schmidt, W.: On solutions of one-dimensional stochastic differential equations without drift. Z. Wahrscheinlichkeitstheor. verw. Geb. **68**, 287–314 (1985)

6. Grigelionis, B.: On generalized z-diffusions. In: Buckdahn, R. et al. (eds.) Stochastic Processes and Related Topics, Stochastic Monographs, Taylor & Francis, pp. 155–169. New - York, London (1998)
7. Crow, J.F., Kimura, M.: An Introduction to Population Genetics Theory. Harper & Row Publishes, New York (1970)
8. Borkovec, M., Klüppelberg, C.: Extremal behaviour of diffusion models in finance. Extremes **1**(1), 47–80 (1998)
9. Hanson, F.B., Tuckwell, H.C.: Diffusion approximation for neural activity including synaptic reversal potentials. J. Theoret. Neurobiol. **2**, 127–153 (1953)
10. Johannesma, P.I.M.: Diffusion models for stochastic activity of neurons. In: Caianiello, E.R. (ed.) Neural Networks, Springer, Berlin (1968)
11. Kallianpur, G., Wolpert, R.: Weak convergence of stochastic neural models. In: Kimura, M., Kallianpur, G., Hida, T. (eds.) Stochastic Methods in Biology, Springer, Berlin (1984)
12. Föllmer, H., Schweizer, M.: A microeconomic approach to diffusion models for stock prices. Math. Financ. **3**(1), 1–23 (1993)
13. Nagahara, Y.: Non-Gaussian distribution for stock returns and related stochastic differential equation. Financ. Eng. Jpn. Markets **3**, 121–149 (1996)
14. Avram, F., Leonenko, N.N., Rabehasaina, L., Šuvak, N.: On ruin theory, Sturm-Liouville theory, spectral decomposition and statistical inference for Wong-Pearson (jump-) diffusions. Working paper (2009)
15. Forman, J.L., Sørensen, M.: The Pearson diffusions: A class of statistically tractable diffusion processes. Scand. J. Stat. **35**(3), 438–465 (2008)
16. Davis, R.A.: Maximum and minimum of one-dimensional diffusions. Stochast. Process. Appl. **13**, 1–9 (1982)
17. Berman, S.M.: Limiting distribution of the maximum of a diffusion. Ann. Math. Stat. **35**, 319–329 (1964)
18. Newell, G.F.: Asymptotic extreme value distribution for one-dimensional diffusion process. J. Math. Mech. **11**, 481–496 (1962)
19. Embrechts, P., Klüppelberg, C., Mikosch, T.: Modelling extremal events for insurance and finance. Springer, Berlin (1997)
20. Leadbetter, M.R., Lindgren, G., Rootzen, H.: Extremes and related properties of random sequences and processes. Springer, Berlin (1983)
21. Grigelionis, B.: An analogue of Gnedenko's theorem for stationary diffusions. Theory Stochast. Process. **8**(24), 1–2, 119–126 (2003)
22. von Mises, R.: La distribution de la plus grande de n valeurs. Revue Mathématique de l'Union Interbalkanique (Athens) **1**, 141–160 (1936)
23. Kallenberg, O.: Random Measures. Akademie Verlag, Berlin (1975)
24. Grigelionis, B.: On point measures of ε-upcrossings for stationary diffusions. Stat. Probab. Lett. **61**, 403–410 (2003)
25. Kolmogorov, A.N.: Über die analytischen. Methoden in der Wahrscheinlichkeitsrechnung. Math. Ann. **104**, 415–458 (1931)
26. Bibby, B.M., Skovgaard, I.M., Sørensen, M.: Diffusion-type models with given marginals and autocorrelation function. Bernoulli **11**, 191–220 (2003)
27. Heyde, C.C., Leonenko, N.N: Student processes, Adv. Appl. Prob. **37**, 342–365 (2005)
28. Kutoyants, Y.A.: Statistical Inference for Ergodic Diffusion Processes. Springer, New York (2004)
29. Kutoyants, Y.A., Yoshida, N.: Moment estimation for ergodic diffusion process. Bernoulli **13**, 933–961 (2007)
30. Leonenko, N.N., Šuvak, N.: Statistical inference for Student diffusion process. Stoch. Anal. Appl. **28**, 972–1002 (2010)
31. Leonenko, N.N., Šuvak, N.: Statistical inference for reciprocal gamma diffusion process. J. Stat. Plan. Infer. **140**, 30–51 (2010)
32. Sørensen, H.: Parametric inference for diffusion processes observed at the discrete points in time. A survey, Int. Stat. Rev. **72**, 337–354 (2004)

33. Abramowitz, M., Stegun, I. (eds.): Handbook of Mathematical Functions. Dover, New York (1968)
34. Titchmarsh, E.C.: Eigenfunctions expansions associated with second order differential equations. Part I, Claredon Press, Oxford (1962)
35. Wong, E.: The construction of a class of stationary Markov processes. In: Belman, R. (ed.) Sixteen Symposium of Applied Mathematics-Stochastic Processes in Mathematical Physics and Engineering, American Mathematics Society, vol. 16, pp. 264–276 (1964)
36. Romanovsky, V.: Sur quelques classes nouvelles de polynomes orthogonaux. Calcul de Probabilités, 1023–1025 (1929)
37. Routh, W.J.: On some properties of certain solutions of a differential equations of the second order. Proc. Lond. Math. Soc. **16**, 245–261 (1885)

Chapter 7
Miscellanea

7.1 Mixed Moments of Student's t-Distributions

Let M_d be the Euclidean space of symmetric $d \times d$ matrices with the scalar product $\langle A_1, A_2 \rangle := \mathrm{tr}(A_1 A_2)$, $A_1, A_2 \in M_d$, $M_d^+ \subset M_d$ be the cone of nonnegative definite matrices and $\mathscr{P}(M_d^+)$ be a class of probability measures on M_d^+. Here $\mathrm{tr}A$ denotes the trace of a matrix A.

The probability distribution of a d-dimensional random vector X is said to be the mixture of centered Gaussian distributions with the mixing distribution $U \in \mathscr{P}(M_d^+)$ (U-mixture for short) if, for all $z \in R^d$,

$$\mathrm{E}e^{i\langle z, X\rangle} = \int\limits_{M_d^+} e^{-\frac{1}{2}\langle zA, z\rangle} U(\mathrm{d}A). \tag{7.1}$$

The distributional properties of such mixtures are well studied (see, e.g., [1, 2] and references therein).

Let $c_j = (c_{j_1}, \ldots, c_{j_d}) \in R^d$, $j = 1, 2, \ldots, 2n$. We shall derive formulas evaluating $\mathrm{E}\left(\prod_{j=1}^{2n} \langle c_j, X\rangle\right)$ for U-mixtures of Gaussian distributions, including Student's t-distribution.

Let Π_{2n} be the class of pairings σ on the set $I_{2n} = \{1, 2, \ldots, 2n\}$, i.e. the partitions of I_{2n} into n disjoint pairs, implying that

$$\mathrm{card}\Pi_{2n} = \frac{(2n)!}{2^n n!}.$$

For each $\sigma \in \Pi_{2n}$, we define uniquely the subsets $I_{2n\backslash\sigma}$ and integers $\sigma(j)$, $j \in I_{2n\backslash\sigma}$, by the equality

$$\sigma = \left\{(j, \sigma(j)), j \in I_{2n\backslash\sigma}\right\}.$$

B. Grigelionis, *Student's* t*-Distribution and Related Stochastic Processes*, SpringerBriefs in Statistics, DOI: 10.1007/978-3-642-31146-8_7,

If $U = \varepsilon_\Sigma$ is a Dirac measure with fixed $\Sigma \in M_d^+$, i.e. the Gaussian case, Isserlis theorem (in mathematical physics known as Wick theorem) says (see, e.g., [3–5]) that

$$\mathrm{E}\left[\prod_{j=1}^{2n} \langle c_j, X\rangle\right] = \sum_{\sigma\in\Pi_{2n}} \prod_{j\in I_{2n\backslash\sigma}} \langle c_j \Sigma, c_{\sigma(j)}\rangle := m_{2n}(c, \Sigma). \tag{7.2}$$

Write

$$\phi_U(\Theta) := \int_{M_d^+} e^{-\mathrm{tr}(A\Theta)} U(\mathrm{d}A), \quad \Theta \in M_d^+. \tag{7.3}$$

Theorem 7.1 *[6] The following statements hold:*

(i) The probability distribution of a d-dimensional random vector X is the U-mixture of centered Gaussian distributions if and only if

$$\mathrm{E}e^{i\langle z,X\rangle} = \phi_U\left(\frac{1}{2}z^T z\right), \tag{7.4}$$

where z^T is the transposed vector z.

(ii) If the probability distribution of X is the U-mixture of centered Gaussian distributions and, for $j = 1, 2, \ldots, 2n$,

$$\int_{M_d^+} \langle c_j A, c_j\rangle^n U(dA) < \infty, \tag{7.5}$$

then

$$\mathrm{E}\left[\prod_{j=1}^{2n} \langle c_j, X\rangle\right] = \sum_{\sigma\in\Pi_{2n}} \int_{M_d^+} m_{2n}^\sigma(c, A) U(dA), \tag{7.6}$$

where

$$m_{2n}^\sigma(c, A) = \prod_{j\in I_{2n\backslash\sigma}} \langle c_j A, c_{\sigma(j)}\rangle.$$

Proof (i) The statement follows from (7.1) and (7.3), because, obviously,

$$\mathrm{tr}\left((z^T z)A\right) = \langle zA, z\rangle.$$

(ii) Observe that card $I_{2n\backslash\sigma} = n$ and, for all $\sigma \in \Pi_{2n}$ and $A \in M_d^+$,

$$
\begin{aligned}
\prod_{j\in I_{2n\setminus\sigma}} \left|\langle c_j A, c_{\sigma(j)}\rangle\right|^n &\le n^{-n}\left(\sum_{j\in I_{2n\setminus\sigma}} \left|\langle c_j A, c_{\sigma(j)}\rangle\right|\right)^n \\
&\le n^{-1}\sum_{j\in I_{2n\setminus\sigma}} \left|\langle c_j A, c_{\sigma(j)}\rangle\right|^n \\
&\le \frac{2^{n-1}}{n}\sum_{j\in I_{2n\setminus\sigma}} \left[\langle c_j A, c_j\rangle^n + \langle c_{\sigma(j)} A, c_{\sigma(j)}\rangle^n\right] \\
&= \frac{2^{n-1}}{n}\sum_{j=1}^{2n}\langle c_j A, c_j\rangle^n. \qquad (7.7)
\end{aligned}
$$

Using (7.5) and (7.7), we find that

$$
\begin{aligned}
\mathrm{E}\left[\prod_{j=1}^{2n}\langle c_j, X\rangle\right] &= \int_{M_d^+} m_{2n}(c, A)U(\mathrm{d}A) \\
&= \sum_{\sigma\in\Pi_{2n}}\int_{M_d^+} m_{2n}^{\sigma}(c, A)U(\mathrm{d}A).
\end{aligned}
$$

□

Taking (see also [7])

$$U = \mathscr{L}(Y\Sigma),$$

where $\Sigma \in M_d^+$ is fixed and

$$\mathscr{L}(Y) = GIG\left(-\frac{\nu}{2}, \nu, 0\right)$$

we have that

$$\phi_U(\Theta) = \frac{2\left(\frac{\nu}{2}\right)^{\frac{\nu}{4}}(\mathrm{tr}(\Sigma\Theta))^{\frac{\nu}{4}}}{\Gamma\left(\frac{\nu}{2}\right)} K_{\frac{\nu}{2}}\left(\sqrt{2\mathrm{tr}(\Sigma\Theta)}\right), \qquad (7.8)$$

$$\mathscr{L}(X) = T_d(\nu, \Sigma, 0) \qquad (7.9)$$

and, for $j = 1, 2, \ldots, 2n$

$$\int_{M_d^+} \langle c_j A, c_j\rangle^n U(\mathrm{d}A) = \begin{cases} \dfrac{\Gamma\left(\frac{\nu}{2} - n\right)}{\left(\frac{\nu}{2}\right)^{\frac{\nu}{2} - n}}\langle c_j\Sigma, c_j\rangle^n, & \text{if} \quad 2n < \nu, \\ \infty, & \text{if} \quad 2n \ge \nu. \end{cases}$$

Thus, for $2n < \nu$,

$$\int\limits_{R^d} \prod_{j=1}^{2n} \langle c_j, x\rangle T_d\,(\nu, \Sigma, 0)\,(\mathrm{d}x) = \frac{\Gamma\left(\frac{\nu}{2} - n\right)}{\left(\frac{\nu}{2}\right)^{\frac{\nu}{2}-n}} m_{2n}(c, \Sigma), \tag{7.10}$$

$$\int\limits_{R^d} \prod_{j=1}^{2n} \langle c_j, x\rangle T_d(\nu, \Sigma, \alpha)(\mathrm{d}x) = \int\limits_{R^d} \prod_{j=1}^{2n} \left[\langle c_j, y\rangle + \langle c_j, \alpha\rangle\right] T_d(\nu, \Sigma, 0)(\mathrm{d}y)$$

and because of anti-symmetry, for $2k + 1 < \nu$,

$$\int\limits_{R^d} \prod_{j=1}^{2k+1} \langle c_j, x\rangle T_d(\nu, \Sigma, 0)(\mathrm{d}x) = 0.$$

Remark 7.2 Let $\nu \geq d$ be an integer, $Y_1, \ldots, Y_\nu$ be i.i.d. d-dimensional centered Gaussian vectors with a covariance matrix Σ, $|\Sigma| > 0$, and $U = \mathscr{L}\left(\nu \Sigma_\nu^{-1}\right)$, where the matrix

$$W_\nu = \sum_{j=1}^{\nu} Y_j^T Y_j.$$

If $\nu \geq d$, the matrix W_ν is invertible with probability 1, because it is well known that the Wishart distribution

$$\mathscr{L}(W_\nu) := W_d(\Sigma, \nu)$$

has a density

$$W_d(\Sigma, \nu, A) = \begin{cases} \dfrac{|A|^{\frac{\nu - d - 1}{2}} \exp\left\{-\frac{1}{2}\mathrm{tr}\left(\Sigma^{-1} A\right)\right\}}{\left(2^d |\Sigma|\right)^{\frac{\nu}{2}} \pi^{\frac{d(d-1)}{4}} \prod\limits_{j=1}^{d} \Gamma\left(\frac{k - j + 1}{2}\right)}, & \text{if } |A| > 0, \\ 0, \quad \text{otherwise.} & \end{cases}$$

Because (see, e.g., [2, 8, 9])

$$\begin{aligned} \int\limits_{M_d^+} e^{-\frac{1}{2}\langle zA, z\rangle} U(\mathrm{d}A) &= \int\limits_{R^d} e^{i\langle z, x\rangle} T_d(\nu, \Sigma, 0)(\mathrm{d}x) \\ &= \mathrm{E}[e^{-\frac{1}{2}\langle z\Sigma, z\rangle Y}], \quad z \in R^d, \end{aligned} \tag{7.11}$$

taking $z = tc, t \in R^1, c \in R^d$, we find that

$$\int_{M_d^+} e^{-\frac{t^2}{2}\langle cA,c\rangle} U(\mathrm{d}A) = \mathrm{E}\left[e^{-\frac{t^2}{2}\langle c\Sigma,c\rangle Y}\right].$$

Thus, for all $c \in R^d$,

$$\mathscr{L}\left(\nu\langle cW^{-1}, c\rangle\right) = \mathscr{L}\left(\langle c\Sigma, c\rangle Y\right),$$

contradicting to the formula

$$\mathscr{L}\left(\langle cW_\nu^{-1}, c\rangle\right) = \mathscr{L}\left(\langle c\Sigma^{-1}, c\rangle \frac{1}{\chi^2_{\nu-d+1}}\right)$$

in [9].

Unfortunately, the last formula was used in [6], Example 3.

From (7.11) we easily find that

$$\int_{R^d} e^{i\langle z,x\rangle} T_d(\nu, \Sigma, \alpha)(\mathrm{d}x) = \frac{e^{i\langle z,\alpha\rangle}}{2^{\frac{\nu}{2}-1}\Gamma\left(\frac{\nu}{2}\right)} \left(\nu\langle z\Sigma, z\rangle\right)^{\frac{\nu}{4}} \times K_{\frac{\nu}{2}}\left(\sqrt{\nu\langle z\Sigma, z\rangle}\right), \quad z \in R^d,$$

(see [10, 11]).

7.2 Long-Range Dependent Stationary Student Processes

It is well known (see, e.g., [12]) that a real square integrable and continuous in quadratic mean stochastic process $X = \{X_t, t \in R^1\}$ is second order stationary if and only if it has the following spectral decomposition:

$$X_t = \alpha + \int_{-\infty}^{\infty} \cos(\lambda t)v(\mathrm{d}\lambda) + \int_{-\infty}^{\infty} \sin(\lambda t)w(\mathrm{d}\lambda), \quad t \in R^1,$$

where $\alpha = \mathrm{E}X_0$, $v(\mathrm{d}\lambda)$ and $w(\mathrm{d}\lambda)$ are mean 0 and square integrable real random measures such that, for each $A, A_1, A_2 \in \mathscr{B}(R^1)$,

$$\mathrm{E}\left[v(A_1)v(A_2)\right] = \mathrm{E}v^2(A_1 \cap A_2), \tag{7.12}$$

$$\mathrm{E}\left[w(A_1)w(A_2)\right] = \mathrm{E}w^2(A_1 \cap A_2), \tag{7.13}$$

$$\mathrm{E}\left[v(A_1)w(A_2)\right] = 0, \tag{7.14}$$

$$\tilde{F}(A) := \mathrm{E}v^2(A) = \mathrm{E}w^2(A). \tag{7.15}$$

The correlation function r satisfies

$$r(t) = \int_{-\infty}^{\infty} \cos(\lambda t) F(\mathrm{d}\lambda),$$

where

$$F(A) = \frac{\tilde{F}(A)}{\tilde{F}(R^1)}, \quad A \in \mathscr{B}(R^1).$$

Following [13], we shall construct a class of strictly stationary stochastic processes $X = \left\{X_t, t \in R^1\right\}$ such that

$$\mathscr{L}(X_t) \equiv T_1\left(\nu, \sigma^2, \alpha\right), \quad \nu > 2,$$

called the Student's stationary processes.

Recall the notion and some properties of the independently scattered random measures (i.s.r.m.) (see [13–15]).

Let $T \in \mathscr{B}(R^d)$, $\mathscr{S}$ be a σ-ring of subsets of T (i.e. countable unions of sets in $\mathscr{S}$ belong to $\mathscr{S}$ and, if $A, B \in \mathscr{S}$, $A \subset B$, then $B \backslash A \in \mathscr{S}$). The σ algebra generated by $\mathscr{S}$ is denoted $\sigma(\mathscr{S})$.

A collection of random variables $v = \{v(A), A \in \mathscr{S}\}$ defined on a probability space $(\Omega, \mathscr{F}, \mathrm{P})$ is said to be an i.s.r.m. if, for every sequence $\{A_n, n \geq 1\}$ of disjoint sets in $\mathscr{S}$, the random variables $v(A_n)$, $n = 1, 2, \ldots$, are independent and

$$v\left(\bigcup_{n=1}^{\infty} A_n\right) = \sum_{n=1}^{\infty} v(A_n) \quad a.s.,$$

whenever $\bigcup_{n=1}^{\infty} A_n \in \mathscr{S}$.

Let $v(A)$, $A \in \mathscr{S}$, be infinitely divisible,

$$\log \mathrm{E}e^{izv(A)} = izm_0(A) - \frac{1}{2}z^2 m_1(A) + \int_{R_0^+} \left(e^{izu} - 1 - iz\tau(u)\right) \Pi(A, \mathrm{d}u),$$

where m_0 is a signed measure, $\Pi(A, \mathrm{d}u)$ for fixed A is a measure on $\mathscr{B}(R_0^1)$ such that

$$\int_{R_0^1} \left(1 \wedge u^2\right) \Pi(A, \mathrm{d}u) < \infty;$$

$$\tau(u) = \begin{cases} u, & \text{if } |u| \le 1, \\ \dfrac{u}{|u|}, & \text{if } |u| > 1. \end{cases}$$

Assume now that $m_0 = m_1 = 0$ and

$$\Pi(A, \mathrm{d}u) = M(A)\Pi(\mathrm{d}u),$$

where $M(A)$ is some measure on T and $\Pi(\mathrm{d}u)$ is some Lévy measure on R_0^1.

Integration of functions on T with respect to ν is defined first for real simple functions $f = \sum_{j=1}^{n} x_j 1_{A_j}$, $A_j \in \mathscr{S}$, $j = 1, \ldots, n$, by

$$\int_A f(x)\nu(\mathrm{d}x) = \sum_{j=1}^{n} x_j \nu(A \cap A_j),$$

where A is any subset of T, for which $A \in \sigma(\mathscr{S})$ and $A \cap A_j \in \mathscr{S}$, $j = 1, \ldots, n$.

In general, a function $f \colon (T, \sigma(\mathscr{S})) \to \left(R^1, \mathscr{B}(R^1)\right)$ is said to be ν-integrable if there exists a sequence $\{f_n, n = 1, 2, \ldots\}$ of simple functions as above such that $f_n \to f$ M-a.e. and, for every $A \in \sigma(\mathscr{S})$, the sequence $\left\{\int_A f_n(x)\nu(\mathrm{d}x),\right.$ $\left. n = 1, 2, \ldots\right\}$ converges in probability, as $n \to \infty$. If f is ν-integrable, we write

$$\int_A f(x)\nu(\mathrm{d}x) = p - \lim_{n\to\infty} \int_A f_n(x)\nu(\mathrm{d}x).$$

The integrand $\int_A f(x)\nu(\mathrm{d}x)$ does not depend on the approximating sequence.

A function f on T is ν-integrable if and only if

$$\int_T Z_0\left(f(x)\right) M(\mathrm{d}x) < \infty$$

and

$$\int_T \left|Z\left(f(x)\right)\right| M(\mathrm{d}x) < \infty,$$

where

$$Z_0(y) = \int_{R_0^1} \left(1 \wedge (uy)^2\right) \Pi(\mathrm{d}u),$$

and

$$Z(y) = \int_{R_0^1} (\tau(uy) - y\tau(u))\, \Pi(\mathrm{d}u).$$

For such functions f

$$\log \mathrm{E} \exp \left\{ i\xi \int_A f(x)\nu(\mathrm{d}x) \right\} = \int_A \varkappa(\xi f(x))\, M(\mathrm{d}x),$$

where

$$\varkappa(\xi) = \int_{R_0^1} \left(e^{i\xi u} - 1 - i\xi\tau(u) \right) \Pi(\mathrm{d}u).$$

Let now $Y_t = \left(Y_t^1, Y_t^2\right)$, $t \geq 0$, be a bivariate Student-Lévy process such that

$$\mathscr{L}(Y_1) = T_2(\nu, \sigma^2 I_2, 0), \quad I_2 = \begin{pmatrix} 1 & 0 \\ 0 & 1 \end{pmatrix},$$

and F be an arbitrary probability distribution on R^1.

Let $T = R^1$, $\mathscr{S}$ be the σ-ring of subsets $A = \bigcup_{j=1}^{\infty} \left(a_j, b_j\right]$, where the intervals $\left(a_j, b_j\right]$, $j = 1, 2, \ldots$, are disjoint. Define i.m.r.m. v and w by the equalities:

$$v(A) = \sum_{j=1}^{\infty} \left(Y^1_{F(b_j)} - Y^1_{F(a_j)} \right)$$

and

$$w(A) = \sum_{j=1}^{\infty} \left(Y^2_{F(b_j)} - Y^2_{F(a_j)} \right), \quad A = \bigcup_{j=1}^{\infty} \left(a_j, b_j\right] \in \mathscr{S}.$$

Because, for $i = 1, 2,\ j = 1, 2, \ldots, \nu > 2$,

$$\mathrm{E}(Y^i_{F(b_j)} - Y^i_{F(a_j)}) = 0,$$

$$\mathrm{E}(Y^i_{F(b_j)} - Y^i_{F(a_j)})^2 = \frac{\sigma^2 \nu}{\nu - 2} \left(F(b_j) - F(a_j) \right)$$

and

$$\sum_{j=1}^{\infty} \mathrm{E}(Y^i_{F(b_j)} - Y^i_{F(a_j)})^2 \leq \frac{\sigma^2 \nu}{\nu - 2} < \infty,$$

the definition of v and w is correct.

From (7.10) it follows that v and w satisfies (7.12)–(7.15) with

$$\tilde{F}(A) = \frac{\sigma^2 \nu}{\nu - 2} F(A), \quad A \in \mathscr{S}.$$

Thus, the process

$$X_t = \alpha + \int_{-\infty}^{\infty} \cos(ut) v(du) + \int_{-\infty}^{\infty} \sin(ut) w(du), \quad t \in R^1,$$

is well defined, strictly stationary,

$$\mathscr{L}(X_t) \equiv T_1(\nu, \sigma^2, \alpha)$$

and the correlation function r satisfies

$$r(t) = \int_{-\infty}^{\infty} \cos(ut) F(du), \quad t \in R^1.$$

Strict stationarity of X follows from the formula (see [13]):

$$\begin{aligned} \mathrm{E} e^{i \sum_{j=1}^{n} \eta_j X_{t_j}} &= e^{i\alpha \sum_{j=1}^{n} \eta_j} \\ &\quad \times \exp\left\{ \int_{-\infty}^{\infty} \log \hat{h}_{\nu,\sigma} \left(\frac{1}{2} \sum_{j,k=1}^{n} \eta_j \eta_k \cos\left(u(t_j - t_k)\right) \right) F(du) \right\}, \\ &\quad \eta_j, t_j \in R^1, \quad j = 1, \ldots, n, \end{aligned}$$

where

$$\begin{aligned} \hat{h}_{\nu,\sigma}(\theta) &:= \int_0^{\infty} e^{-\theta u} \frac{1}{\sigma^2} gig\left(\frac{u}{\sigma^2}; -\frac{\nu}{2}, \nu, 0 \right) du \\ &= \frac{2}{\Gamma\left(\frac{\nu}{2}\right)} \left(\frac{\theta \sigma^2 \nu}{2} \right)^{\frac{\nu}{4}} K_{\frac{\nu}{2}}\left(\sqrt{2\sigma^2 \theta \nu} \right), \quad \theta > 0. \end{aligned}$$

As it was checked in [16], if

$$F(du) = f_{\beta,\gamma}(u) du, \quad 0 < \beta \leq 1, \quad \gamma \in R^1,$$

where

$$f_{\beta,\gamma}(u) = \frac{1}{2}\left[f_{\beta,0}(u+\gamma) + f_{\beta,0}(u-\gamma)\right], \quad u \in R^1,$$

with

$$f_{\beta,0}(u) = \frac{2^{\frac{1-\beta}{2}}}{\sqrt{\pi}\Gamma\left(\frac{\beta}{2}\right)} K_{1-\beta}(|u|)\,|u|^{\frac{(1-\beta)}{2}},$$

then

$$r(t) = \frac{\cos\gamma t}{(1+t^2)^{\frac{\beta}{2}}}, \quad t \in R^1,$$

and

$$\int_{-\infty}^{\infty} |r(t)|\,\mathrm{d}t = \infty,$$

implying long-range dependence of X (see also [17–20]).

Remark 7.3 Defining Student-Lamperti process $X^\star$ as (see [21])

$$X_t^\star = t^H X_{\log t}, \quad t > 0, \quad X_0^\star = 0, \quad H > 0.$$

we have that $X^\star$ is H-self-similar, i.e., for each $c > 0$, processes $\left\{X_{ct}^\star, t \geq 0\right\}$ and $\left\{c^H X_t^\star, t \geq 0\right\}$ have the same finite dimensional distributions, and (see [13])

$$\mathrm{E}e^{i\sum_{j=1}^{n}\eta_j X_{t_j}^\star} = e^{i\alpha\sum_{j=1}^{n} t_j^H \eta_j} \times \exp\left\{\int_{-\infty}^{\infty}\left[\log\hat{h}_{\nu,\sigma}\left(\frac{1}{2}\sum_{j,k=1}^{n}\eta_j\eta_k t_j^H t_k^H \cos\left(u\log\frac{t_j}{t_k}\right)\right)\right]F(\mathrm{d}u)\right\},$$

$$t_j > 0, \quad \eta_j \in R^1, \quad j = 1,\ldots,n.$$

In particular,

$$\mathrm{E}e^{i\eta X_t^\star} = e^{i\alpha t^H\eta}\hat{h}_{\nu,\sigma}\left(t^{2H}\frac{\eta^2}{2}\right), \quad t > 0, \quad \eta \in R^1,$$

and

$$\mathrm{E}e^{i\eta(X_t^\star - X_s^\star)} = e^{i\alpha(t^H - s^H)\eta}\exp\left\{\int_{-\infty}^{\infty}\left[\log\hat{h}_{\nu,\sigma}\left(\frac{1}{2}\eta^2\left(s^{2H} + t^{2H} - 2s^H t^H\cos\left(u\log\frac{t}{s}\right)\right)\right)\right]F(\mathrm{d}u)\right\}, \quad s,t > 0, \quad \eta \in R^1.$$

7.3 Lévy Copulas

Considering the probability distributions F on R^d with the 1-dimensional Student's t marginals $F_{j,j} = 1, \ldots, d$, and having in mind their relationship with stochastic processes, we restricted ourselves to the cases when F is a mixture of the d-dimensional Gaussian distributions .

Denoting

$$C(u_1, \ldots, u_d) := F\left(F_1^{-1}(u_1), \ldots, F_d^{-1}(u_d)\right), \quad u_j \in [0, 1], \quad j = 1, \ldots, d,$$

it is obvious that this function is the probability distribution function on the d-cube $[0,1]^d$ with uniform one-dimensional marginals, called the d-copula (see, e.g., [22]). Trivially,

$$F(x_1, \ldots, x_d) = C\left(F_1(x_1), \ldots, F_d(x_d)\right), \quad (x_1, \ldots, x_d) \in R^d. \tag{7.16}$$

Formula (7.16) with the arbitrary d-copula defines uniquely the probability distributions on R^d with the given Student's 1-dimensional marginals. These statements are very special cases of well known Sklar's theorem (see [23, 24]).

Thus, taking concrete d-copulas we shall obtain a wide class of multivariate generalizations of Student's t-distributions.

For instance, the Archimedean copulas have the from

$$C(u_1, \ldots, u_d) = \psi\left(\psi^{-1}(u_1) + \cdot + \psi^{-1}(u_d)\right), \quad u_j \in [0, 1], \quad j = 1, \ldots, d,$$

where ψ is a d-monotone function on $[0, \infty)$, i.e., for each $x \geq 0$ and $k = 0, 1, \ldots, d-2$,

$$(-1)^k \frac{\mathrm{d}^k}{\mathrm{d}x^k} \psi(x) \geq 0,$$

$(-1)^{d-2}\psi^{(d-2)}(x)$, $x \geq 0$, is nonincreasing and convex function.

In particular, if

$$\psi(x) = (1+x)^{-\frac{1}{\theta}}, \quad \theta \in (0, \infty), \quad x \geq 0,$$

we have the Clayton's copula

$$C(u_1, \ldots, u_d) = \left(\sum_{j=1}^{d} u_j^{-\theta} - d + 1\right)^{-\frac{1}{\theta}}, \quad u_j \in [0, 1], \quad j = 1, \ldots, d.$$

If $\phi(x) = \exp\left\{-x^{\frac{1}{\theta}}\right\}$, $\theta \geq 1$, $x \geq 0$, we obtain the Gumbel copula

$$C(u_1,\dots,u_d)=\exp\left\{-\left(\sum_{j=1}^{d}(-\log u_j)^{\theta}\right)^{\frac{1}{\theta}}\right\},\quad u_j\in[0,1],\quad j=1,\dots,d.$$

Unfortunately, it is difficult to describe if the copulation preserves such important for us properties of marginal distributions as infinite divisibility or self-decomposability.

A promising direction for future work is a notion of Lévy copulas and, analogously to the classical copulas, construction of new Lévy measures on R^d using marginal ones (see [25–28]). Following [28], we briefly describe an analogue of Sklar's theorem in this context.

Let $\bar{R}:=(-\infty,\infty]$. For $a,b\in\bar{R}^d$ we write $a\le b$, if $a_k\le b_k$, $k=1,\dots,d$ and, in this case, denote

$$(a,b]:=(a_1,b_1]\times\ldots\times(a_d,b_d].$$

Let $F:S\to\bar{R}$ for some subset $S\subset\bar{R}^d$. For $a,b\in S$ with $a\le b$ and $\overline{(a,b]}\subset S$, the F-volume of $(a,b]$ is defined by

$$V_F((a,b]):=\sum_{u\in\{a_1,b_1\}\times\cdots\times\{a_d,b_d\}}(-1)^{N(u)}F(u),$$

where $N(u):=\sharp\{k:u_k=a_k\}$.

A function $F:S\to\bar{R}$ is called d-increasing if $V_F((a,b])\ge 0$ for all $a,b\in S$ with $a\le b$ and $\overline{(a,b]}\subset S$.

Definition 7.4 Let $F:\bar{R}^d\to\bar{R}$ be a d-increasing function such that $F(u_1,\dots,u_d)=0$ if $u_i=0$ for at least one $i\in\{1,\dots,d\}$. For any non-empty index set $I\subset\{1,\dots,d\}$ the I-marginal of F is the function $F_I:\bar{R}^{|I|}\to\bar{R}$, defined by

$$F^I((u)i)_{i\in I}):=\lim_{a\to\infty}\sum_{(u_i)_{i\in I^c}\in\{-a,\infty\}^{|I^c|}}F(u_1,\dots,u_d)\prod_{i\in I^c}\mathrm{sgn}u_i,$$

where $I^c=\{1,\dots,d\}\backslash I$, $|I|:=\mathrm{card}I$, and

$$\mathrm{sgn}x=\begin{cases}1, & \text{if } x\ge 0,\\ -1, & \text{if } x<0.\end{cases}$$

Definition 7.5 A function $F:\bar{R}^d\to\bar{R}$ is called a Lévy copula if

1. $F(u_1,\dots,u_d)\ne\infty$ for $(u_1,\dots,u_d)\ne(\infty,\dots,\infty)$,
2. $F(u_1,\dots,u_d)=0$ if $u_i=0$ for at least one $i\in\{1,\dots,d\}$,
3. F is d-increasing,
4. $F^{\{i\}}(u)=u$ for any $i\in\{1,\dots,d\}$, $u\in R^1$.

Write

$$\mathscr{I}(x) := \begin{cases} (x, \infty), & \text{if } x \le 0, \\ (-\infty, x], & \text{if } x > 0. \end{cases}$$

Definition 7.6 Let $X = (X^1, \ldots, X^d)$ be an R^d-valued Lévy process with the Lévy measure Π. The tail integral of X is the function $V : \left(R^1 \backslash \{0\}\right)^d \to R^1$ defined by

$$V(x_1, \ldots, x_d) := \prod_{i=1}^{d} \operatorname{sgn}(x_i) \Pi\left(\mathscr{I}(x_1) \times \cdots \times \mathscr{I}(x_d)\right)$$

and, for any non-empty $I \subset \{1, \ldots, d\}$ the I-marginal tail integral V^I of X is the tail integral of the process $X^I := (X^i)_{i \in I}$.

We denote one-dimensional margins by $V_i := V^{\{i\}}$.

Observe, that marginal tail integrals $\{V^I \colon I \subset \{1, \ldots, d\} \text{ non-empty}\}$ are uniquely determined by Π. Conversely, Π is uniquely determined by the set of its marginal tail integral.

Relationship between Lévy copulas and Lévy processes are described by the following analogue of Sklar's theorem.

Theorem 7.7 [28]

1. *Let $X = \left(X^1, \ldots, X^d\right)$ be an R^d-valued Lévy process. Then there exists a Lévy copula F such that the tail integrals of X satisfy*

$$V\left((x_i)_{i \in I}\right) = F^I\left((V_i(x_i))_{i \in I}\right), \tag{7.17}$$

   for any non-empty $I \subset \{1, \ldots, d\}$ and any $(x_i)_{i \in I} \in \left(R^1 \backslash \{0\}\right)^{|I|}$. The Lévy copula F is unique on $\operatorname{Ran} V_1 \times \cdots \times \operatorname{Ran} V_d$.

2. *Let F be a d-dimensional Lévy copula and V_i, $i = 1, \ldots, d$, be tail integrals of real-valued Lévy processes. Then there exists an R^d-valued Lévy process X whose components have tail integrals $V_1, \ldots, V_d$ and whose marginal tail integrals satisfy (7.17) for any non-empty $I \subset \{1, \ldots, d\}$ and any $(x_i)_{i \in I} \in \left(R^1 \backslash \{0\}\right)^{|I|}$. The Lévy measure Π of X is uniquely determined by F and V_i, $i = 1, \ldots, d$.*

In the above formulation $\operatorname{Ran} V$ means the range of V. The reader is referred for proofs to [28].

An analogue of the Archimedean copulas is as follows (see [28]).

Let $\varphi : [-1, 1] \to [-\infty, \infty]$ be a strictly increasing continuous function with $\varphi(1) = \infty$, $\varphi(0) = 0$, and $\varphi(-1) = -\infty$, having derivatives of orders up to d on $(-1, 0)$ and $(0, 1)$, and, for any $k = 1, \ldots, d$, satisfying

$$\frac{\mathrm{d}^k \varphi(u)}{\mathrm{d}u^k} \geq 0, \quad u \in (0,1) \quad \text{and} \quad (-1)^k \frac{\mathrm{d}^k \varphi(u)}{\mathrm{d}u^k} \leq 0, \quad u \in (-1,0).$$

Let

$$\tilde{\varphi}(u) := 2^{d-2} \left(\varphi(u) - \varphi(-u)\right), \quad u \in [-1,1].$$

Then

$$F(u_1, \dots, u_d) := \varphi\left(\prod_{i=1}^{d} \tilde{\varphi}^{-1}(u_i)\right)$$

defines a Lévy copula.

In particular, if

$$\varphi(x) := \eta \left(-\log|x|\right)^{-\frac{1}{\vartheta}} 1_{\{x>0\}} - (1-\eta)\left(-\log|x|\right)^{-\frac{1}{\vartheta}} 1_{\{x<0\}}$$

with $\vartheta > 0$ and $\eta \in (0,1)$, then

$$\tilde{\varphi}(x) = 2^{d-2} \left(-\log|x|\right)^{-\frac{1}{\vartheta}} \operatorname{sgn} x, \quad x \in -1, 1],$$

and

$$F(u_1, \dots, u_d) = 2^{2-d} \left(\sum_{i=1}^{d} |u_i|^{-\vartheta}\right)^{-\frac{1}{\vartheta}} \left(\eta 1_{\{u_1 \dots u_d \geq 0\}} - (1-\eta) 1_{\{u_1 \dots u_d < 0\}}\right),$$

resembling the ordinary Clayton copulas.

References

1. Barndorff-Niesen, O.E., Pedersen, J., Sato, K.: Multivariate subordination, self-decomposability and stability. Adv. Prob. **33**, 160–187 (2001)
2. Barndorff-Niesen, O.E., Perez-Abreu, V.: Extensions of type G and marginal infinite divisibility. Teor. Veroyatnost. Primenen. **47**(2), 301–319 (2002)
3. Isserlis, L.: On a formula for the product-moment coefficient of any order of a normal frequency distribution in any number of variables. Biometrika **12**, 134–139 (1918)
4. Leonov, V.P., Shiryaev, A.N.: An a method of calculation of semi-invariants. Theory Probab. Appl. **4**, 319–329 (1959)
5. Simon, B.: The $P(\Phi)_2$ euclidean (quantum) field theory. Princeton University Press, New Jersy, (1974)
6. Grigelionis, B.: On the Wick theorem for mixtures of centered Gaussian distribution. Lith. Math. J. **49**(4), 372–380 (2009)
7. Vignat, C., Bhatnagar, S.: An extension of Wick's theorem. Stat. Probab. Lett. **78**(15), 2400–2403 (2008)
8. Johnson, N.L., Kotz, S.: Distributions in Statistics: continuous univariate distributions. Wiley, New York (1972)

9. Rao, C.R.: Linear statistical inference and its applications, Wiley, NewYork (1965)
10. Heyde, C.C., Leonenko, N.N.: Student processes, Adv. Appl. Prob. **37**, 342–365 (2005)
11. Joarder, A.H., Ali, M.M.: On the characteristic function of the multivariate t-distribution. Pak. J. Stat. **12**, 55–62 (1996)
12. Cramér, H., Leadbetter, M.R.: Stationary and Related Stochastic Processes. Wiley, New York (1967)
13. Barndorff-Niesen, O.E., Perez-Abreu, V.: Stationary and self-similar processes driven by Lévy processes. Stoch. Process. Appl. **84**, 357–369 (1999)
14. Kwapień, S., Woyozyński, N.A.: Random Series and Stochastic Integrals: single and multiple. Birkhäuser, Boston (1992)
15. Rajput, B.S., Rosinski, J.: Spectral representations of infinitely divisible processes. Probab. Theory Relat. Fields **82**, 451–487 (1989)
16. Anh, V.V., Knopova, V.P., Leonenko, N.N.: Continuous-time stochastic processes with cyclical long-range dependence. Austral. N. Z. J. Statist. **46** 275–296 (2004)
17. Anh, V.V., Heyde, C.C., Leonenko, N.N.: Dynamic models of long-memory processes driven by Lévy noise. J. Appl. Probab. **39**, 730–747 (2002)
18. Barndorff-Nielsen, O.E.: Superpositions of Ornstein-Uhlenbeck processes. Theory Probab. Appl. **45**(2), 175–194 (2001)
19. Barndorff-Niesen, O.E., Leonenko, N.N.: Spectral properties of superpositions of Ornstein-Uhlenbeck type processes. Method. Comput. Appl. Prob. **7**(3), 335–352 (2005)
20. Heyde, C.C., Yang, Y.: On defining long-range dependence. J. Appl. Prob. **34**, 939–944 (1997)
21. Lamperti, J.W.: Semi-stable stochastic processes. Trans. Am. Math. Soc. **104**, 62–78 (1962)
22. Nelsen, R.B.: An Introduction to Copulas. Springer, New York (1999)
23. Sklar, A.: Fonctions de répartition à n dimensions et leurs margés. Pub. Inst. Stat. Univ. Paris **8**, 229–231 (1959)
24. Sklar, A.: Random variables, distribution functions, and copulas-a personal look backward and forward. In: Distributions with fixed marginals and related topics. In: Rüschendorf, L., Schweizer, B., Taylor, M.D. (eds.), Institute of Mathematical Statistics, Hayward (1996)
25. Barndorff-Niesen, O.E., Lindner, A.M.: Lévy copulas: dynamics and transforms of upsilon type. Scand. J. Stat. **34**(2), 298–316 (2007)
26. Cont, R., Tankov, P.: Financial Modelling with Jump Processes. Chapman & Hall/CRC, Boca Raton (2004)
27. Jaworski, P., Durante, F., Handle, F., Rychlik, T. (eds.): Copula theory and its applications. Lecture notes in statistics, Springer (2010)
28. Kallsen, J., Tankov, P.: Characterization of dependence of multidimensional Lévy processes using Lévy copulas. J. Multivar. Anal. **97**, 1551–1572 (2006)

Appendix A
Bessel Functions

Bessel functions of the first kind $J_{\pm\nu}(z)$, of the second kind$Y_\nu(z)$ and of the third kind $H_\nu^{(1)}(z)$ and $H_\nu^{(2)}(z)$ are solutions of the differential equation:

$$z^2\frac{\mathrm{d}^2w}{\mathrm{d}z^2}+z\frac{\mathrm{d}w}{\mathrm{d}z}+(z^2-\nu^2)w=0.$$

The function $J_\nu(z)$ can be represented as the following series:

$$J_\nu(z)=\sum_{m=0}^{\infty}\frac{(-1)^m\left(\frac{1}{2}z\right)^{\nu+2m}}{m!\Gamma(m+\nu+1)},\qquad |\arg z|<\pi,$$

$$Y_\nu(z)=\frac{J_\nu(z)\cos(\nu\pi)-J_{-\nu}(z)}{\sin(\nu\pi)},$$

where the right-hand side of the last equation is replaced by its limiting value if ν is an integer or zero,

$$H_\nu^{(1)}(z)=J_\nu(z)+iY_\nu(z)=\frac{1}{i\sin(\nu\pi)}\left(J_{-\nu}(z)-J_\nu(z)e^{-i\nu\pi}\right),$$

$$H_\nu^{(2)}(z)=J_\nu(z)-iY_\nu(z)=\frac{1}{\sin(\nu\pi)}\left[J_\nu(z)e^{i\nu\pi}-J_{-\nu}(z)\right].$$

Modified Bessel functions of the first kind

$$I_\nu(z)=\begin{cases}e^{-i\nu\frac{\pi}{2}}J_\nu\left(e^{i\frac{\pi}{2}}z\right), & -\pi<\arg z\le\frac{\pi}{2},\\ e^{-3i\nu\frac{\pi}{2}}J_\nu\left(e^{-3i\frac{\pi}{2}}z\right), & \frac{\pi}{2}<\arg z\le\pi,\end{cases}$$

B. Grigelionis, *Student's t-Distribution and Related Stochastic Processes*,
SpringerBriefs in Statistics, DOI: 10.1007/978-3-642-31146-8,

and of the third kind

$$K_\nu(z) = \frac{1}{2} i\pi e^{i\pi\frac{\nu}{2}} H_\nu^{(1)}\left(e^{i\frac{\pi}{2}} z\right)$$
$$= -\frac{1}{2} i\pi e^{-i\pi\frac{\nu}{2}} H_\nu^{(2)}\left(e^{-i\frac{\pi}{2}} z\right)$$

and satisfies the formulas:

$$I_\nu(z) = \sum_{m=0}^{\infty} \frac{\left(\frac{1}{2}z\right)^{\nu+2m}}{m!\Gamma(m+\nu+1)}, \quad \nu > -1,$$

$$K_\nu(z) = \frac{\pi}{2} \frac{I_\nu(z) - I_{-\nu}(z)}{\sin(\nu\pi)},$$

where the right hand side of the last equation is replaced by its limiting value if ν is an integer or zero.

When $\nu = n + \frac{1}{2}$, $n = 0, 1, \ldots$, the Bessel functions are elementary:

$$J_{n+\frac{1}{2}}(z) = \sqrt{\frac{2}{z}} z^{n+\frac{1}{2}} \left(-\frac{1}{z}\frac{\mathrm{d}}{\mathrm{d}z}\right)^n \frac{\sin z}{z},$$

$$J_{-n-\frac{1}{2}}(z) = \sqrt{\frac{2}{\pi}} z^{n+\frac{1}{2}} \left(\frac{1}{z}\frac{\mathrm{d}}{\mathrm{d}z}\right)^n \frac{\cos z}{z},$$

$$Y_{n+\frac{1}{2}}(z) = (-1)^{n+1} J_{-n-\frac{1}{2}}(z),$$

$$K_{n+\frac{1}{2}}(z) = (-1)^n \sqrt{\frac{\pi}{2z}} z^{n+1} \left(\frac{\mathrm{d}}{z\mathrm{d}z}\right)^n \frac{e^{-z}}{z},$$

$$I_{n+\frac{1}{2}}(z) = \sqrt{\frac{2}{\pi}} z^{n+\frac{1}{2}} \left(\frac{\mathrm{d}}{z\mathrm{d}z}\right)^n \frac{\sinh z}{z}.$$

The following integral representations and useful formulas hold true:

$$K_\nu(z) = K_{-\nu}(z) = \frac{1}{2}\left(\frac{z}{2}\right)^\nu \int_0^\infty t^{-\nu-1} e^{-t-\frac{z^2}{4t}} \mathrm{d}t$$
$$= \frac{1}{2}\int_0^\infty t^{-\nu-1} \exp\left\{-\frac{1}{2}z\left(t+t^{-1}\right)\right\} \mathrm{d}t, \quad z > 0,$$

$$K_{\nu+1}(z) = \frac{2\nu}{z} K_\nu(z) + K_{\nu-1}(z),$$

$$K_{\nu+1}(z) + K_{\nu-1}(z) = -2K_\nu'(z),$$

$$z^{\nu}K_{\nu}(z) = \int_{z}^{\infty} t^{\nu}K_{\nu-1}(t)\mathrm{d}t,$$

$$K_{\nu-1}(z) = zK_{\nu}(z)\int_{0}^{\infty}\frac{g_{\nu}(t)}{z^2+t}\mathrm{d}t \quad \text{(Grosswald's formula)},$$

where

$$g_{\nu}(t) = 2\left\{\pi^2 t\left(J_{\nu}^2(\sqrt{t}) + Y_{\nu}^2(\sqrt{t})\right)\right\}^{-1}, \quad t > 0,$$

$$K_{\nu}(z) \sim \sqrt{\frac{\pi}{2z}}e^{-z}, \quad \text{as } z \to \infty,$$

$$K_0(z) \sim \ln\frac{1}{z}, \quad \text{as } z \to 0,$$

$$z^{|\nu|}K_{\nu}(z) \uparrow \Gamma(|\nu|)2^{|\nu|-1}, \quad \text{as } z \downarrow 0, \quad (\nu \neq 0),$$

$$K_{\nu}(\nu z) \sim \sqrt{\frac{\pi}{2\nu}}\frac{e^{-\nu\sqrt{1+z^2}}}{(1+z^2)^{\frac{1}{4}}}\left(\frac{z}{1+\sqrt{1+z^2}}\right)^{-\nu}, \quad \text{as} \quad \nu \to \infty,$$

uniformly with respect to real z.

Index

B. Grigelionis, *Student's* t-*Distribution and Related Stochastic Processes*,
SpringerBriefs in Statistics, DOI: 10.1007/978-3-642-31146-8,